Perspectives on Development in the Middle East and North Africa (MENA) Region

Series editor

Almas Heshmati, Sogang University, Seoul, Korea (Republic of)

This book series publishes monographs and edited volumes devoted to studies on the political, economic and social developments of the Middle East and North Africa (MENA). Volumes cover in-depth analyses of individual countries, regions, cases and comparative studies, and they include both a specific and a general focus on the latest advances of the various aspects of development. It provides a platform for researchers globally to carry out rigorous economic, social and political analyses, to promote, share, and discuss current quantitative and analytical work on issues, findings and perspectives in various areas of economics and development of the MENA region. Perspectives on Development in the Middle East and North Africa (MENA) Region allows for a deeper appreciation of the various past, present, and future issues around MENA's development with high quality, peer reviewed contributions. The topics may include, but not limited to: economics and business, natural resources, governance, politics, security and international relations, gender, culture, religion and society, economics and social development, reconstruction, and Jewish, Islamic, Arab, Iranian, Israeli, Kurdish and Turkish studies. Volumes published in the series will be important reading offering an original approach along theoretical lines supported empirically for researchers and students, as well as consultants and policy makers, interested in the development of the MENA region.

More information about this series at http://www.springer.com/series/13870

Masoomeh Rashidghalam

Measurement and Analysis of Performance of Industrial Crop Production: The Case of Iran's Cotton and Sugar Beet Production

Masoomeh Rashidghalam
Department of Agricultural Economics
University of Tabriz
Tabriz
Iran

ISSN 2520-1239   ISSN 2520-1247   (electronic)
Perspectives on Development in the Middle East and North Africa (MENA) Region
ISBN 978-981-13-0091-2   ISBN 978-981-13-0092-9   (eBook)
https://doi.org/10.1007/978-981-13-0092-9

Library of Congress Control Number: 2018938645

Printed on acid-free paper

This Springer imprint is published by the registered company Springer Nature Singapore Pte Ltd.
part of Springer Nature
The registered company address is: 152 Beach Road, #21-01/04 Gateway East, Singapore 189721, Singapore

To my parents, Mohammad and Zari

Acknowledgements

I would like to express my sincere thanks to many people and universities that have helped me to achieve my academic goals and who deserve my acknowledgment.

I am thankful to Prof. Ghader Dashti, Dr. Esmaeil Pishbahar and all my professors at Department of Agricultural Economics, University of Tabriz, for providing me financial support and scientific formation during all my studies.

I would also like to thank Prof. Almas Heshmati for his valuable suggestions and contributions to this book. The hospitability of the Jönköping International Business School (JIBS), Jönköping University and Sogang University, faculty, staff and graduate students is highly appreciated.

I am deeply and forever indebted to all my family for their tremendous love, continued support and encouragement.

Acknowledgements

Contents

Abbreviations

AE	Allocative efficiency
AR	Autocorrelation
BCC	Banker, Charnes and Cooper
CCR	Charnes, Cooper and Rhodes
COLS	Corrected ordinary least squares
CRS	Constant returns to scale
DEA	Data envelopment analysis
DFA	Distribution-free analysis
DMU	Decision-making units
EE	Economic efficiency
FDH	Free disposal hull
FE	Fixed effects
GLS	Generalized least squares
ML	Maximum likelihood
MLE	Maximum likelihood estimation
MOLS	Modified ordinary least squares
OLS	Ordinary least squares
OTE	Overall technical efficiency
PFA	Partial frontier analysis
PTE	Persistent technical efficiency
RE	Random effect
RTE	Residual technical efficiency
SFA	Stochastic frontier analysis
SF-Panel	Stochastic frontier panel
TE	Technical efficiency
TFA	Thick frontier analysis
VRS	Variable returns to scale

List of Figures

List of Tables

Chapter 1
Introduction

Abstract This chapter gives a general introduction of the book. It provides a perspective on the important role that the agricultural sector plays in developing societies' growth and development. This sector is needed for increasing crop production so that countries can meet the food and clothing needs of their increasing populations. Therefore, policy priorities for the agriculture sector include a quantitative analysis of production for increasing crop production. In this regard, Iran needs to increase the domestic production of industrial crops in two possible ways: first, by increasing the area under cultivation and second, increasing yield per unit of land. Due to limited supply of arable land and production inputs, the first alternative is not practical. Therefore, increasing industrial crop production from the land already under cultivation seems to be a better solution. In other words, if Iran can increase the amount of Industrial crop produced from one hectare of cultivated area, its total production will increase. This chapter also provides the main goals, assumptions and questions of the study that the rest of the chapters address.

1.1 Introduction

One of the most important aims of developing countries is achieving high levels of economic growth. Any country that increases its productive capacity can improve its society's welfare and it can also compete in international markets. Among the different ways in which production can be increased in developing countries are increasing the use of production factors. However, this faces many problems and limitations. In this regard, increasing technical efficiency is considered as an appropriate solution. Increasing technical efficiency leads to higher labor productivity, it can create more production from a fixed set of production factors and it is particularly important for avoiding resource waste.

In developing countries the agricultural sector is one of the most active and productive sectors of the economy. Traditionally, a large share of the population in these countries has been engaged in agriculture related activities. In addition to employment opportunities, this sector also provides access to food and enhanced food and

© Springer Nature Singapore Pte Ltd. 2018   1
M. Rashidghalam, *Measurement and Analysis of Performance of Industrial Crop Production:*
The Case of Iran's Cotton and Sugar Beet Production, Perspectives on Development in the Middle
East and North Africa (MENA) Region, https://doi.org/10.1007/978-981-13-0092-9_1

textile security. In these countries the agricultural sector is the producer of essential goods and given their growing populations the demand for demand for agricultural products is increasing considerably. On the other hand, agricultural products are also main items of exports and earning foreign exchange. Therefore, this sector also has an important role to play in increasing these countries' GDP. Since agricultural production is a function of basic resources or factors of production such as land, labor, capital and technology, and due to the limitations of these production factors, it is important to increase technical efficiency, that is, increase production while maintaining the *same* level of *inputs*.

This book does a productivity and efficiency analysis to identify product enhancements while maintaining and conserving natural resources. It also uses this as a complement to the policies adopted by the Iranian government.

Iran is considered a developing country which faces main problems of lack of sufficient levels of production and low economic growth. One of the main reasons for this is the existing low level of technical efficiency and inadequate use of production factors, in particular in the agricultural sector. Achieving high production levels and higher economic growth are not possible without increasing physical capacity and consumption of inputs. If these cannot be increased, then the only alternative is increasing technical efficiency. One also needs appropriate information for proper national and regional planning including knowledge about the different regions' capacities. Hence, the positioning of different regions, especially in terms of unused economic capacities and facilities for increasing production, has great importance for increasing land and labor productivity. Success here requires paying attention to the potential of different provinces when national and regional development plans are being formulated. This involves identifying the capabilities of different provinces in terms of optimal allocation of resources and the proper use of inputs and various other economic factors (Amadeh et al. 2009).

1.2 Problem Statement

Man's economic efforts have always been focused on the goal of maximizing output using the lowest amount of inputs or cost minimization of production for a given level of output. This can be called achievement of higher efficiency and productivity. Productivity is a comprehensive concept that encompasses efficiency and is regarded by policymakers and economists as a way of improving the standard of living, well-being and tranquility in society (Abrishami and Niakan 2010). In Iran, with its limited production factors and low yield in its agricultural sector, the most appropriate way of achieving and appreciating its growth rate appears to be by improving the performance and efficiency of producers.

Industrial crops refer to a set of crops that are not specifically grown for food like fruits, vegetables and grains are, but are specifically grown to yield useful products for human or industrial processes. Products from some industrial crops can be used directly or without any industrial processes either directly or indirectly by humans or livestock (Khajepour 1994). Given the importance of industrial crops and the significant role that they play in Iran's agricultural economics, this book analyzes the technical efficiency of different sugar beet and cotton producing provinces in the country. These two commodities are used in textile, sugar and animal food industries.

Cotton is one of the most valuable industrial crops in Iran. It is also one of the most important and strategic agricultural products that has been cultivated for a long time. In some provinces the climatic and land conditions are suitable for growing cotton. As an essential ingredient in the textile industry, cotton has great importance in the economic status of the country. Different cotton products reflect the capacity of this industrial crop in creating job opportunities in the agricultural and industrial sectors thus confirming its important role in economic development (Haeri Asayesh 2009).

With a 115-year production history, sugar beet is an industrial and strategic product in Iran. This crop is also of central importance because of population growth and the increasing need for sugar. Since sugar beet molasses are used in feeding livestock, it has an important role in the household food basket, in conversion and in livestock industries. Government policy for farming and producing sugar beet is based on extensive interference in its imports to keep sugar prices low to meet the needs of local consumers. Because a large part of produced sugar beet is used only in the sugar industry, most of the active factories have started cultivating and producing sugar beet (Yazdani and Rahimi 2012). Today, sugar beet and sugar cane are responsible for 40 and 60% of the total produced sugar respectively (Seyyed Sharifi 2013). In total, sugar beet is cultivated by 95,000 farmers in Iran.

Provinces' performance in producing cotton and sugar beet is given in Tables 1.1 and 1.2 respectively. As seen in these tables, there is a significant difference between the performances of provinces producing these products which can be related to environmental inequalities, climatic conditions and technical efficiency. Therefore, paying attention to the efficiency measure in the regional dimension for developing agricultural activities for reducing some regional problems, especially inequalities between the regions, can be beneficial. Besides, improving crop efficiency is an important and effective factor in increasing the productivity of production factors and their performance without imposing any additional costs. This makes paying attention to improving efficiency at both the micro and macroeconomic levels (national) of central importance.

As seen in Table 1.1, mean cotton production yield in Iran is about 2 tons per hectare. However, worldwide on average farmers harvest 6 tons per hectare which shows that the yield of Iranian cotton is about one-third of the world level (Ministry of Agriculture Jihad 2015). By considering these factors and the importance of industrial crops in the country's economy and due to the reduction of these products in Iran, this book analyzes their technical efficiency in different provinces and evaluates effective factors of the technical efficiency of producing these products.

Table 1.1 Area harvested, production, and yield of cotton producing provinces per hectare in 2012–2013 (unit: hectare-ton-kg)

Province	Area harvested			Production			Yield	
	Irrigated	Rainfed	Total	Irrigated	Rainfed	Total	Irrigated	Rainfed
East Azerbaijan	800	–	800	1720	–	1720	2149	–
Ardabil	1355	–	1355	2864	–	2864	2114	–
Isfahan	2300	–	2300	5307	–	5307	2190	–
Tehran	240	–	240	516	–	516	2150	–
South Khorasan	9501	–	9501	19874	–	19874	2091	–
Razavi Khorasan	27800	–	27800	59235	–	59235	2130	–
North Khorasan	7500	–	7500	16523	–	16523	2203	–
Semnan	2550	–	2550	5703	–	5703	2236	–
Fars	14550	–	14550	35223	–	35223	2426	–
Qom	1810	–	1810	3878	–	3878	2142	–
Kerman	940	–	940	2102	–	2102	2235	–
Golestan	11199	1520	12719	23394	2140	25534	2088	1408
Mazandaran	1150	380	1530	2379	530	2909	2069	1394
Markazi	380	–	380	872	–	872	2294	–
Yazd	100	–	100	256	–	256	2558	–

Source: Ministry of Agriculture Jihad (2015)

Table 1.2 Area harvested, production and yield of sugar beet producing provinces per hectare in 2012–2013 (unit: hectare-ton-kg)

Province	Area harvested			Production			Yield	
	Irrigated	Rainfed	Total	Irrigated	Rainfed	Total	Irrigated	Rainfed
West Azerbaijan	31113	–	31113	1835000	–	1835000	58978	–
Isfahan	1500	–	1500	53755	–	53755	35836	–
Chaharmahal and Bakhtiari	1204	–	1204	47700	–	47700	39617	–
South Khorasan	830	–	830	26600	–	26600	32056	–
Razavi Khorasan	16500	–	16500	716369	–	716369	43415	–
North Khorasan	4088	–	4088	161650	–	161650	39538	–
Semnan	2541	–	2541	115001	–	115001	45252	–
Fars	10818	–	10818	503314	–	503314	46524	–
Qazvin	2700	–	2700	92000	–	92000	334069	–
Kermanshah	9000	–	9000	425133	–	425133	47236	–
Lorestan	4355	–	4355	202041	–	202041	46394	–
Markazi	841	–	841	33500	–	33500	39814	–
Whole country	97101	–	97101	473995	–	473995	48722	–

Source: Ministry of Agriculture Jihad (2015)

1.3   Importance and Necessity of Research on Cotton and Sugar Beet Sectors

Iran has suitable conditions and talent for producing industrial crops like cotton and sugar beet which have significant importance because of the large area on which they are cultivated. These products are used in sectors like the general and textile industry, food industry and animal food husbandry. They not only provide raw materials for the textile and oil industries but also play an important role in employment generation in the agriculture, industry and commerce sectors. Therefore, increasing the production of these crops as strategic products is significant for Iran's self-sufficiency. This is important as Iran has been subject to frequent sanctions by western powers.

The average annual consumption of cotton in the last five years was more than 100,000 tons in the textile industry; annually about 50,000 tons of cotton were imported from Central Asia. These amounts are related to the fact that the textile industry is not working to its their full capacity otherwise the minimum annual requirement of cotton fiber is estimated to be about 150,000 tons. With the annual consumption of cotton-based clothes increasing worldwide the aim of cotton producing countries is creating domestic value-added and avoiding importing cotton as raw material. Besides, increasing cotton production especially when there is limited availability of water resources in the world worsens the problem.

Cotton producers in Central Asia have long term plans for developing process industries domestically. For example, Uzbekistan which now supplies 90% of the cotton for spinning industries in Iran aims to develop its own textile and clothing industries; therefore, it is likely to have no surplus cotton for export in the next 10 years and all the cotton will turn to domestic value added. Other cotton producing countries too are following similar strategies.

Iran can take a few measures to improve its cotton production and reduce its dependence on imports. Among these are increasing land under cultivation, improving yield per unit area and increasing the producers' efficiency. The first approach has a certain limitation and recently the area under cotton cultivation has been reduced significantly due to different reasons including lack of competitiveness with other products, high costs of production and harvesting and lack of suitable cultivars showing that this approach is not suitable for increasing production. Hence, improving yields per unit area and increasing producers' efficiency become more appropriate approaches to follow.

The second industrial crop which plays an important role in Iran's food security is sugar beet which this book discusses in detail. As a main energy source sugar plays an important role in supplying calories to the people. In Iran, sugar is produced using sugar beet, sugar cane, imported sugar and processed food. Iran's annual sugar consumption per capita is about 25–28 kg which seems to be increasing because of improvements in people's welfare and rapid population growth. In Iran, sugar is mainly obtained from sugar beet. Many studies that have investigated the changing procedures in the area under sugar beet cultivation indicate a significant reduction in the area under sugar beet cultivation in the last few years. Therefore, to achieve self-sufficiency in sugar production what is needed is not only more investments in the sugar industry but also more facilities being provided to increase sugar beet production and productivity (Rashidghalam 2008).

In Iran, total production capacity and facilities are not fully utilized in the agricultural sector. Therefore, all studies on inefficiency in producing agricultural products discuss ways of improving yields and achieving optimal use of resources to increase agricultural productivity. Considering the facilities and existing limitations in Iran's agricultural sector, it can be said that the most appropriate solution for increasing production and incomes is through properly applying the existing production factors and improving efficiency in the use of the factors through proper management. Therefore, increasing yield per unit area can be a proper approach. It should be paid special attention for increasing efficiency in the agricultural sector, particularly for sugar beet as it is an important agricultural product.

Considering the importance of an efficiency-oriented approach in the economy to analyze economic units (provinces) with regard to their outputs and inputs and solving the main problems faced by this study of lack of coherent research based on a comprehensive approach and efficiency-based provincial divisions in the agricultural sector, this book evaluates the technical efficiency of each province for cotton and sugar beet products during 2000–2012. Based on this it determines the most efficient provinces in producing each product. Besides, it also identifies the technical efficiency determinants of the provinces. Measuring technical efficiency becomes significant as some resources face lack of resources and facilities.

This research evaluates the performance and efficiency in producing sugar beet and cotton in each province in Iran. The findings of this book will help researchers and policymakers to get a clearer idea of the technical efficiency of each province in the production of cotton and sugar beet. The book discusses three significant aspects: measuring the technical efficiency of each province based on available data, ranking the provinces and providing solutions for improving efficiency levels in the production of industrial products. Accordingly, the results can be beneficial for cotton and sugar beet production and consequentially for economic growth.

Hence, the three main contributions of this book are: first, it fills the gap between a methodological analysis of panel data stochastic frontier models and its empirical application in the agricultural sector by reviewing existing stochastic panel data models in literature. Second, it considers the effect of time in non-parametric models using a Window analysis. Time effect is usually ignored in non-parametric frontier models, especially in new partial frontier models (for example, Order-m and Order-α models). Third, the book compares different parametric and non-parametric models and also chooses the models which are best suited for explaining the variations in the datasets based on different test procedures.

The main purpose of this book is evaluating the parametric and non-parametric approaches in measuring the technical efficiency of industrial crops. Its secondary goals are:

1. Estimating frontier production functions for cotton and sugar beet
2. Non-parametric modeling of technical efficiency using a Window Analysis

3. Identifying the determinants of technical efficiency based on parametric models
4. Ranking cotton and sugar beet producing provinces and identifying the most efficient provinces using parametric and non-parametric models
5. Measuring the technical efficiency of the provinces and its changes during the study period

It does this in different chapters which address the following questions:

1. How much technical efficiency exists in the production of industrial crops?
2. Are there any differences in the technical efficiency of provinces in industrial crop production?
3. How do technical efficiency determinants such as 'ratio of chemical fertilizers to total fertilizers' and 'percentage employing machinery' affect provinces' efficiency?
4. What are the differences between the non-parametric models in terms of technical efficiency and ranking of provinces?
5. Which provinces are the most efficient in the production of industrial crops?
6. How did technical efficiency develop in each province during 2000–2012?

1.4 Outline

The main focus of this book is on different parametric and non-parametric models to measure technical efficiency of Iranian provinces in cotton and sugar beet production. It assesses the differences between the models based on characteristics and efficiency score measurements using a systematic sensitivity analysis of the results. The next chapter, therefore introduces the status of agricultural industrial crops in Iran with a focus on cotton and sugar beet production. This chapter also provides an up-to-date picture of the import and export status of these crops for Iran. In addition, it provides an evaluation of the productivity of cotton and sugar beet production in the country. Chapter 3 includes the theoretical framework of efficiency measurement and review the literature of efficiency measurement from the early parametric panel data models to new and more recent models. Chapter 4 gives a description of the twelve parametric and four non-parametric models that the book uses for its analysis. This chapter also describes the application of a Window analysis in non-parametric models. Chapter 5 discusses models for cotton and sugar beet producing provinces in Iran. This chapter includes 26 models measuring technical efficiency and provides a comparison of the different models. Based on these findings Chap. 6 suggests policy recommendations for the policymakers and further research. Finally, Chap. 7 presents some concluding comments, along with a summary of the empirical findings.

Chapter 2
Industrial Crop Production

Abstract This chapter provides detailed information about the area cultivated, production and productivity of cotton and sugar beet in Iran by discussing the production and productivity of these two crops in different provinces of the country. It compares these numbers with the rest of world. It states that one of the main problems in Iran is lack of production and little economic growth. Low level of technical efficiency in production and inadequate use of the factors of production are important reasons for this state of affairs. Achieving higher production and higher economic growth is not possible without increasing physical capacity and consumption of inputs except through increased efficiency. Therefore, improving efficiency both at the micro and macro levels (national) is of great importance.

2.1 Status of Crops in Iran (by Product)

This book studies crops including the main crops produced on a relatively large scale and include cereals, legumes, industrial crops, vegetables, vine, forage and other products.

2.1.1 Area Harvested

During the 2015–2016 crop year, 11.77 million hectares of harvested area in Iran included 8.44 million hectares for cereals (71.75%), 787,000 hectares for legumes (6.96%), 490,000 hectares for industrial crops (4.17%), 530,000 hectares for vegetables (4.51%), 326,000 hectares for vine (2.77%), 1.05 million hectares for forage (8.92%) and 140,000 hectares for other crops (1.19%) (Ministry of Agriculture Jihad 2017). Figure 2.1 shows the percentage of area under different cropsharvested in Iran during the 2015–2016 crop year. Cereals were grown in the largest crop area.

As seen in Fig. 2.1, the largest harvested area belonged to the cereals group (71.75%) followed by forage (8.92%) and legumes (6.69%). These three groups

© Springer Nature Singapore Pte Ltd. 2018

M. Rashidghalam, *Measurement and Analysis of Performance of Industrial Crop Production: The Case of Iran's Cotton and Sugar Beet Production*, Perspectives on Development in the Middle East and North Africa (MENA) Region, https://doi.org/10.1007/978-981-13-0092-9_2

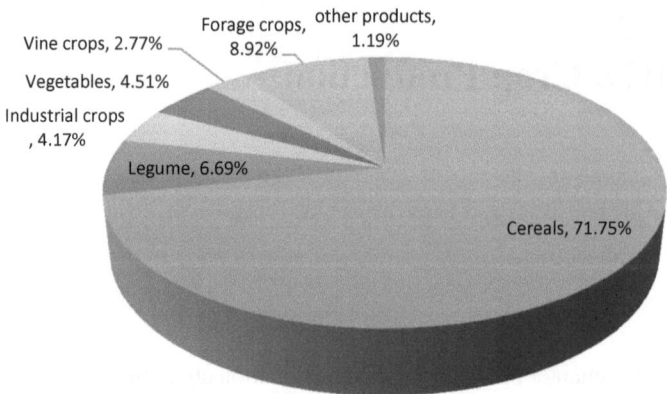

Fig. 2.1   Percentage of crop area harvested in Iran during the 2015–2016 crop year

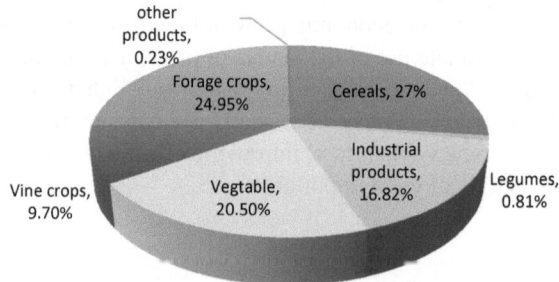

Fig. 2.2   Percentage of crop production in Iran during the 2015–2016 crop year

accounted for 87.37% of the total harvested area. The share of cereals harvested was lower than its share of cultivated area suggesting lower productivity. Traditional agriculture and low rate of use of fertilizers and irrigation induce use of crop rotation to achieve land fertility in a natural way. This rotation is often based on crop root penetration to use different depths of the soil in different years.

2.1.2   Production

The 83 million tons of crops produced in the 2015–2016 crop year included about 22.41 million tons of cereals (27%), about 670,000 tons of legumes (0.81%), about 13.96 million tons of industrial crops(16.82%), about 17.01 million tons of vegetables (20.5%), about 8.05 million tons of vine (9.7%), about 20.7 million tons of forage (24.95%) and about 188,7000 tons of other crops (0.23%). Percentage of production of different crops is given in Fig. 2.2.

As seen in Fig. 2.2, the highest production rate was for the cereals group (27%) followed by forage (24.95%) and vegetables (20.5%). These three groups accounted for 72.45% of the total crop production. The lowest production rate was for other products (0.23%) and legumes (0.81%). A large population and dependency on traditional agriculture and food combined with a lengthy period of international sanctions led to an emphasis on the security of food supply and its variations over time and locations.

2.2 Status of Industrial Crops in Iran

This section discusses the common types of crops grown in Iran. These include industrial crops cotton, tobacco, sugar beet, sugar cane, soybean, rapeseed and other oilseeds. The main focus of this book is on cotton and sugar beet though it also discusses the other crops.

2.2.1 Area Harvested

Based on statistics published by the Ministry of Agriculture Jihad (2015), during the 2015–2016 crop year, industrial crops with 490.200 hectares of harvested area accounted for 4.17% of the harvested area in Iran. It should be noted that 91.6 and 8.4 percent of 490.2000 hectares respectively were cultivated using irrigated and rainfed farming. According to Fig. 2.3, sugar beet (22.5%), sugar cane (18.3%), cotton (14.4%), soybean (10.6%) and rapeseed (10.66%) were responsible for 76.5% of the total area under industrial crops and were ranked first to fifth respectively. Moreover, according to Fig. 2.3, tobacco, sesame, safflower, oilseed sunflower and other oilseeds accounted for 2.2, 8.7, 1.3, 2.5 and 8.8% of the total area under industrial crops harvested respectively.

Fig. 2.3 Percentage distribution of area harvested under industrial crops during the 2015–2016 crop year

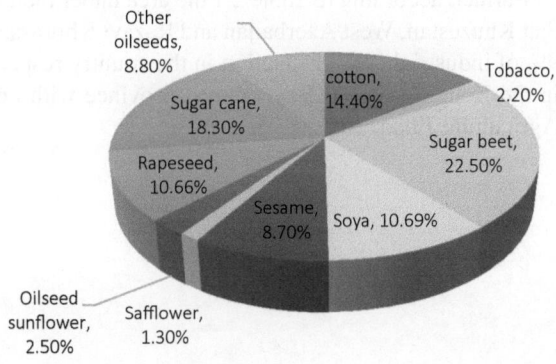

Fig. 2.4 Percentage
distribution of industrial crop
production in Iran during the
2015–2016 crop year

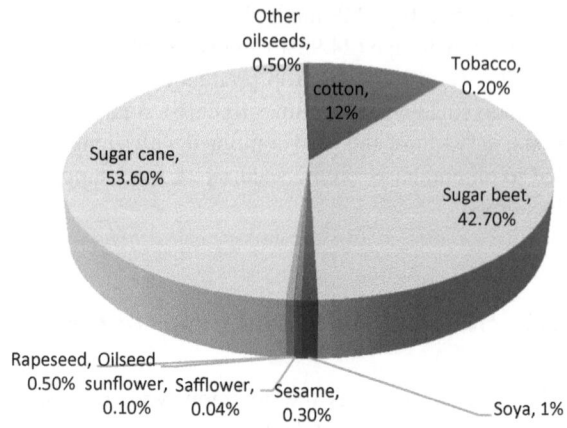

The area allocated for industrial crops across provinces in Table 2.1 shows that Khuzestan, West Azerbaijan, Razavi Khorasan, Fars and Kermanshah provinces accounted for 23.4, 11.9, 11.7, 11.4 and 8.6 of the total industrial crop area in the country respectively. The smallest area was in Kohgiluyeh and Boyerahmad province with only 280 hectares.

2.2.2 Production

In the 2015–2016 crop year, 16.82% (about 14 million tons) of the total crop production was in industrial crops in which 91.6 and 0.3% respectively was cultivated using irrigated and rainfed farming technologies. As seen in Fig. 2.4, sugar cane and sugar beet, which accounted for 53.6 and 42.7% of the industrial crop production respectively, held first and second ranks. This means that about 96.3% of the industrial crop production was of sugar cane and sugar beet (Ministry of Agriculture Jihad 2017) (Fig. 2.5).

Further, according to Table 2.1 the area under industrial crops by province shows that Khuzestan, West Azerbaijan and Razavi Khorasan accounted for 56.2, 13.8 and 9% of industrial crop production in the country respectively. The smallest area was in the Kohgiluyeh and Boyerahmad province with only 280 hectares (Ministry of Agriculture Jihad, 2017) (Fig. 2.6).

Table 2.1 Distribution of area harvested and production of industrial crops during the 2015–2016 crop year (unit: ton-hectare)

Rank	Province	Area harvested		Production	
		Quantity	Percentage	Quantity	Percentage
1	Khuzestan	114500	23.4	7839658	56.2
2	West Azerbaijan	58095	11.9	1931293	13.8
3	Razavi Khorasan	57532	11.7	1259655	9.0
4	Fars	55773	11.4	346826	4.5
5	Kermanshah	42235	8.6	600051	4.3
6	Hamedan	26088	5.3	328005	2.3
7	Lorestan	15862	3.2	262194	1.9
8	Ardabil	13319	2.7	189926	1.4
9	Semnan	12713	2.6	141157	1.0
10	Golestan	8905	1.8	130047	0.9
11	Qazvin	8528	1.7	118667	0.9
12	North Khorasan	8487	1.7	105426	0.8
13	Isfahan	7735	1.6	103935	0.7
14	Kordestan	7703	1.6	63076	0.5
15	Chaharmahal and Bakhtiari	7635	1.6	61750	0.4
16	South Khorasan	7439	1.5	43596	0.3
17	Markazi	6294	1.3	27352	0.2
18	Mazandaran	5100	1.0	24872	0.2
19	East Azerbaijan	3762	0.8	220443	0.3
20	Ilam	3343	0.7	21743	0.2
21	Southof Kerman	3181	0.6	11449	0.1
22	Bushehr	2470	0.5	7996	0.1
23	Hormozgan	2450	0.5	7122	–
24	Gilan	2310	0.5	5440	–
25	Qom	1814	0.4	4695	–
26	Kerman	1761	0.4	3545	–
27	Yazd	1655	0.3	3345	–
28	Zanjan	1640	0.3	2866	–
29	Sistanand Baluchestan	861	0.2	1571	–
30	Alborz	376	0.1	647	–
31	Tehran	317	0.1	471	–
32	Kohgiluyeand Boyerahmad	280	0.1	206	–
–	Whole country	490163	100	13959210	100

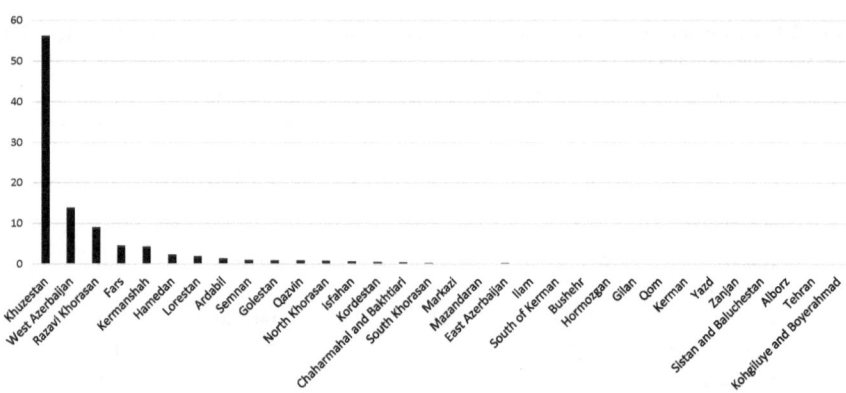

Fig. 2.5 Percentage distribution of industrial crops production in different provinces during the 2015–2016 crop year

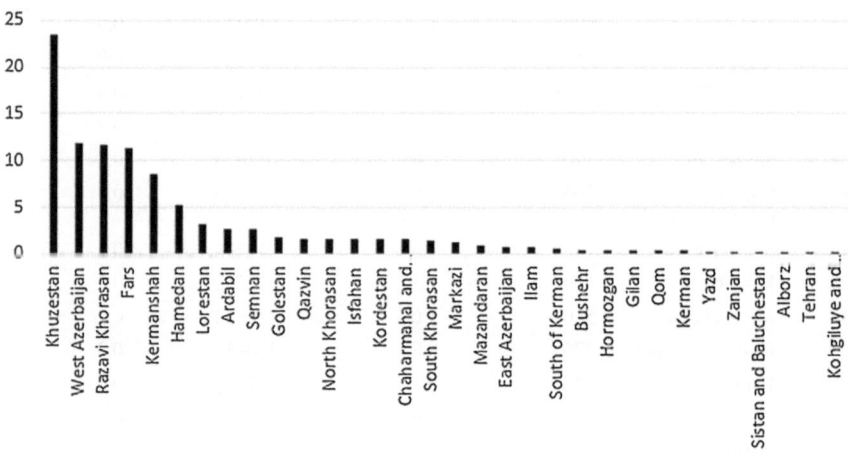

Fig. 2.6 Percentage distribution of area harvested under industrial crops in different provinces during the 2015–2016 crop year

2.3 Status of Cotton Production in Iran

2.3.1 Area Harvested

Based on statistics published by the Ministry of Agriculture Jihad, in the 2014–2015 crop year, cotton with 72,000 hectares of area (equivalent to 0.63% of the total area under crops) accounted for 16.62% of the total industrial crop area in Iran. It should be noted that 99.7% of this land was cultivated using irrigated farming.

Figure 2.7 shows the percentage distribution of cotton cultivated area in Iran's provinces in the 2014–2015 crop year. Based on this figure, Razavi Khorasan

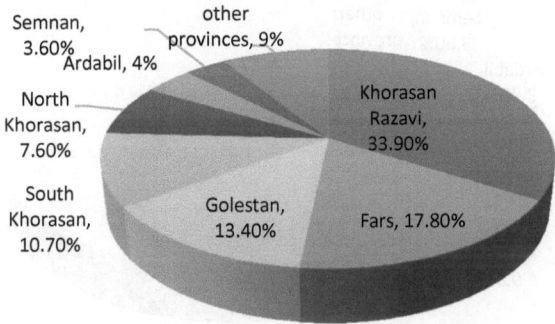

Fig. 2.7 Percentage distribution of area harvested under cotton crop in Iran's provinces in the 2014–2015 crop year

(33.9%), Fars (17.79%), Golestan (13.39%), South Khorasan (10.66%) and North Khorasan (7.63%) were ranked first to fifth cotton producing provinces and accounted for 76.5% of the total cotton harvested area in the country. The lowest cotton production with a harvested area of 11 hectares (0.01) was in Tehran province.

2.3.2 *Production*

Iran's total cotton production was estimated to be about 175,000 million tons that accounted for 0.23 and 1.3% of the total production of all crops and total production of industrial crops respectively; 99.78% of cotton production in the country used irrigated farming.

Figure 2.8 shows the percentage distribution of cotton production by each province in 2014–2015. Razavi Khorasan and Fars, accounted for 33.02 and 20.52% of cotton production respectively and stood in the first and second distribution ranges. This means that about 53.54% of cotton production was in these two provinces. Golestan, South Khorasan and North Khorasan with 10.4, 9.3, and 7.61% cotton production respectively were ranked in the third to fifth positions in cotton production. The lowest cotton production was in Tehran province at only 19 tons (0.01% of the total production in the year).

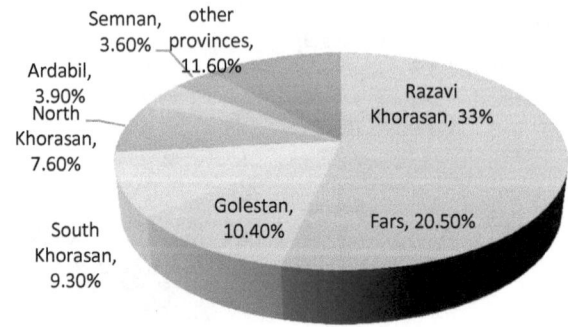

Fig. 2.8 Percentage distribution of cotton crop production in Iran's provinces in the 2014–2015 crop year

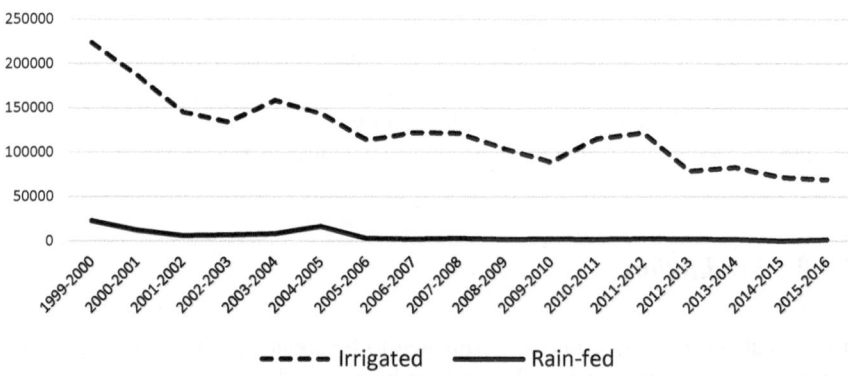

Fig. 2.9 Cotton area harvested during 2000–2016 (unit: hectares)

2.3.3 Yield

In Iran, the yield of cotton in irrigated and rainfed farming was 2,445 and 1,786 kg per hectare respectively. Qazvin and West Azerbaijan provinces with 4,839 and 700 kg/ha yield had the highest and lowest irrigated cotton yield respectively. Gholestan and Mazandaran with 1,802 and 1,443 kg/ha yield were the only provinces that produced cotton using rainfed farming technology (Table 2.2).

Table 2.3 and Figs. 2.9 and 2.10 show the harvested area and production of cotton during 2000–2016. According to these figures, both these decreased during this period. From the 2000 crop year to the 2016 crop year the harvested area reduced by 69%, which means that 223,498 hectares at the beginning of the period reduced to 69,102 hectares at the end of the period. Production decreased from 469,000 tons to 159,000 tons despite a 9% increase in yield (Ministry of Agriculture Jihad 2017).

Table 2.2 Area harvested, production and yield per hectare of cotton crop by province during the 2014–2015 crop year (unit: hectare-ton-kg)

Province	Area harvested			Production			Yield	
	Irrigated	Rainfed	Total	Irrigated	Rainfed	Total	Irrigated	Rainfed
East Azerbaijan	1122	–	1122	3511	–	3511	3128.3	–
West Azerbaijan	35	–	35	24	–	24	700.0	–
Ardabil	2844	–	2844	6894	–	6894	2423.9	–
Isfahan	949	–	949	3188	–	3188	3358.2	–
Alborz	143	–	143	555	–	555	3883.3	–
Tehran	11	–	11	19	–	19	1810	–
South Khorasan	7660	–	7660	16311	–	16311	2129	–
Razavi Khorasan	24352	–	24352	57935	–	57935	2379	–
North Khorasan	5481	–	5481	13354	–	13354	2436	–
Semnan	2602	–	2602	6361	–	6361	2444	–
Fars	12777	–	12777	36010	–	36010	2818	–
Qazvin	281	–	281	1359	–	1359	4839	–
Qom	1301	–	1301	3683	–	3683	2831	–
Kerman	1242	–	1242	3683	–	3683	2966	–
Gholestan	96614	206	9408	18241	372	17869	1899	1802
Mazandaran	79	10	69	131	14	117	1699	1443
Markazi	656	–	656	2022	–	2022	3081	–
Hormozgan	600	–	600	2033	–	2033	3389	–
Yazd	81	–	81	144	–	144	1779	–
Whole country	71828	216	71612	175456	386	175070	2444	1785

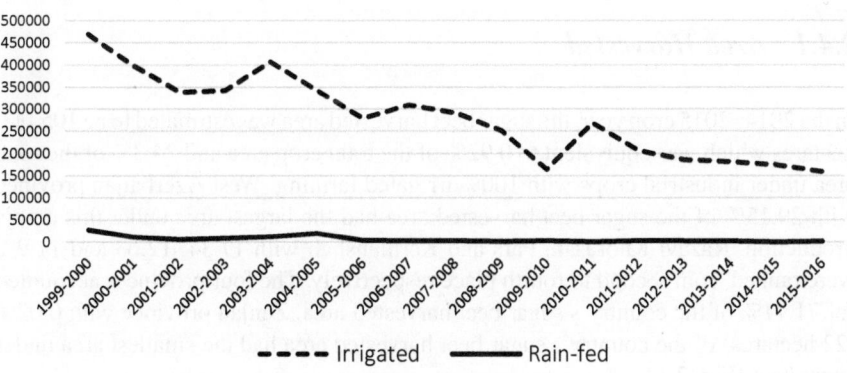

Fig. 2.10 Cotton production during 2000–2016 (unit: tons)

Figure 2.10 shows the production rate of cotton during 2000–2016. According to this figure, 2000 and 2004 crop years had the highest production levels. The

Table 2.3 Area harvested, production and yield per hectare of cotton crop during 2000-16 (unit: hectare-ton-kg)

Year	Area harvested			Production			Yield	
	Irrigated	Rainfed	Total	Irrigated	Rainfed	Total	Irrigated	Rainfed
1999–2000	223498	22728	246226	469050	28074	497124	2098.68	1235.22
2000–2001	186640	12185	198825	399243	12336	411579	2139.11	1012.39
2001–2002	145236	5985	151221	339307	6129	345436	2336.25	1024.06
2002–2003	133500	6587	140087	342533	9243	351776	2565.79	1403.22
2003–2004	158442	8231	166673	407121	13104	420225	2569.53	1592.03
2004–2005	143233	16291	159524	343589	19871	363460	2398.81	1219.75
2005–2006	113345	3215	116560	279337	4335	283672	2464.48	1348.37
2006–2007	122027	2497	124524	309364	3602	312966	2535.21	1442.53
2007–2008	121353	3079	124432	291941	3743	295684	2405.72	1215.65
2008–2009	103539	1831	105370	251635	1969	253604	2430.34	1075.37
2009–2010	88793	2226	91019	164182	3244	167426	1849.04	1457.32
2010–2011	115182	1947	117129	269434	1376	270810	2339.20	706.73
2011–2012	121932	2651	124583	206686	3314	210000	1695.09	1250.09
2012–2013	78718	2398	81116	186561	3034	189595	2369.99	1265.22
2013–2014	82899	1899	84799	181327	2670	183997	2187.32	1406.00
2014–2015	71612	216	71828	175070	386	175456	2444.70	1787.04
2015–2016	69102	1523	70625	159075	2088	161163	2302.03	1370.98

production of cotton declined during the last years. The lowest amount of cotton production was in 2010 and 2016.

2.4 Status of Sugar Beet Production in Iran

2.4.1 Area Harvested

In the 2014–2015 crop year, the sugar beet harvested area was estimated to be 105,000 hectares which was equivalent to 0.92% of the total crop area and 24.3% of the total area under industrial crops with 100% irrigated farming. West Azerbaijan province with 29.15% of the sugar beet harvested area had the largest area under this crop's production. Razavi Khorasan, Fars and Kermanshah with 17.34, 12.63 and 11.9% were ranked from second to fourth place respectively. The four provinces accounted for 71.11% of the country's sugar beet harvested area. Zanjan province with 0.02% (22 hectares) of the country's sugar beet harvested area had the smallest area under sugar beet (Fig. 2.11).

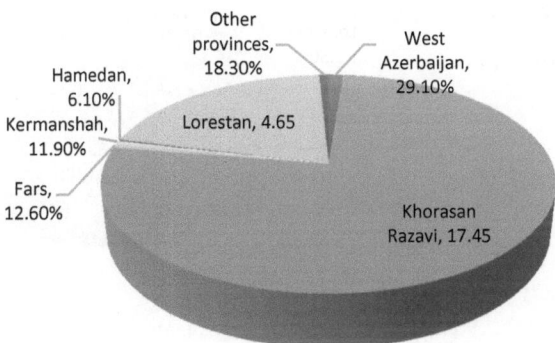

Fig. 2.11 Percentage distribution of area harvested under sugar beet in Iran's provinces in the 2014–2015 crop year

2.4.2 *Production*

The production of sugar beet in Iran was estimated to be approximately 5.6 million tons during the 2014–2015 crop year which was equivalent to 7.26% of the total production of crops and 41.56% of the total production of industrial crops. West Azarbaijan province with 33.14% production took the first place in sugar beet production. Because of suitable soil, availability of water and use of single node seeds this was of the top producers in the country for high-quality products. The use of improved cultivars, mechanized cultivation, high technical knowledge of sugar beet producers in the region and timely educational and informational activities of the Jihad-e-Agriculture Organization are the main reasons for the superiority of West Azarbaijan in sugar beet production.

The failure to pay sugar beet growers' payment claims, lack of financial powers and banks' inability to pay for working capital facilities are some of the most important problems for sugar beet producers in West Azarbaijan province. Despite significantly higher productivity, several studies have shown that the traditional method of sugar beet farming, lack of financial resources, farmers' low level of technical knowledge and drought are some of the problems faced by sugar beet producers in West Azarbaijan (Ministry of Agriculture Jihad, 2016). Razavi Khorasan, Fars and Kermanshah with 16.45, 11.78 and 11.17% of the production respectively were ranked from second to fourth positions in sugar beet production in Iran. These provinces accounted for 72.54% of the country's sugar beet production. At 550 tons (0.01%), the lowest sugar beet production was in Zanjan province (Fig. 2.12).

Fig. 2.12   Percentage distribution of sugar beet production in Iran's provinces in the 2014–2015 crop year

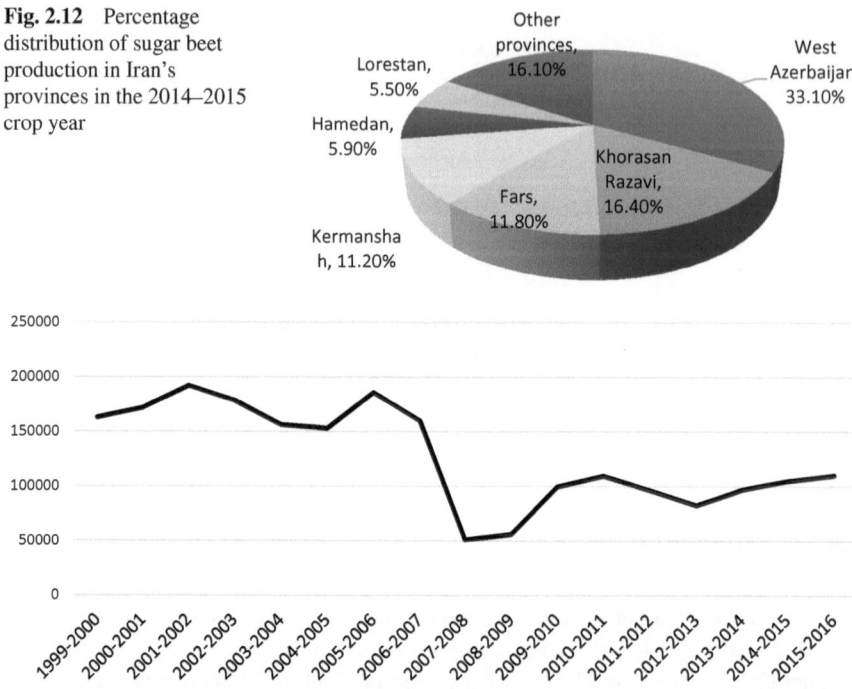

Fig. 2.13   Sugar beet area harvested during 2000–2016 (unit: hectares)

2.4.3   Yield

In Iran, the yield of sugar beet was 53,260 kg per hectare. Khuzestan and Zanjan provinces with 69,572 and 25,000 kg/ha yield had the highest and lowest sugar beet yields (Ministry of Agriculture Jihad, 2016) (Table 2.4).

Figures 2.13 and 2.14 and Table 2.5 present the harvested area and production of sugar beet during the 2000–2016 production years. As seen from these figures and the table, sugar beet average yield per hectare had an increasing trend during this period. A study of the harvested area and production indicates that during this period there were many fluctuations and during 2000–2016 the harvested area varied between 51,000 hectares (in 2008) and 191,000 hectares (in 2002). Despite the increasing trend of sugar beet production in Iran, a systematic model of sugar beet yields is not possible. The yield of this crop ranged between 26 tons per hectare (in 2000) and 54 tons per hectare (in 2016) (see Table 2.5).

Table 2.4   Area harvested, production and yield per hectare of sugar beet crop by province during the 2014-15 crop year (unit: hectare-ton-kg)

Province	Area harvested			Production			Yield	
	Irrigated	Rainfed	Total	Irrigated	Rainfed	Total	Irrigated	Rainfed
East Azerbaijan	120	–	120	6500	–	6500	53986	–
West Azerbaijan	30613	–	30613	1853867	–	1853867	60557.8	–
Ardabil	2743	–	2743	147000	–	147000	53590.9	–
Isfahan	1642	–	1642	58609	–	58609	35696	–
Ilam	203	–	203	9849	–	9849	48470	–
Chaharmahal and Bakhtiari	1314	–	1314	51000	–	51000	38818	–
South Khorasan	929	–	929	34450	–	34450	37102	–
Razavi Khorasan	18305	–	18305	920004	–	920004	50206	–
North Khorasan	2239	–	2239	84500	–	84500	37748	–
Khuzestaan	4442	–	4442	309046	–	309046	69572	–
Zanjan	22	–	22	550	–	550	25000	–
Semnan	2000	–	2000	92506	–	92506	46260	–
Fars	13271	–	13271	659002	–	659002	49658	–
Qazvin	1973	–	1973	82367	–	82367	41749	–
Kordestan	945	–	945	43614	–	43614	46137	–
Kermanshah	12500	–	12500	624999	–	624999	49999	–
Gholestan	66	–	66	2270	–	2270	34498	–
Lorestan	4800	–	4800	260000	–	260000	54168	–
Markazi	544	–	544	25000	–	25000	45938	–
Hamedan	6366	–	6366	329107	–	329107	51695	–
Whole country	105036	–	105036	5594240	–	5594240	53260	–

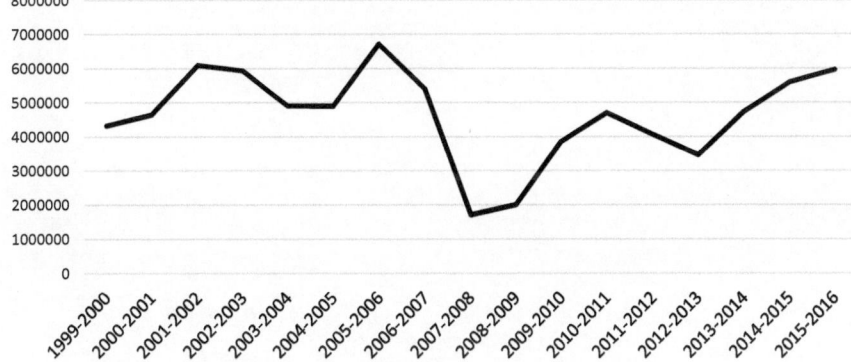

Fig. 2.14   Sugar beet production during 2000–2016 (unit: tons)

Table 2.5 Area harvested, production and yield per hectare of sugar beet crop during 2000-16 (unit: hectare-ton-kg)

Year	Area harvested			Production			Yield	
	Irrigated	Rainfed	Total	Irrigated	Rainfed	Total	Irrigated	Rainfed
1999–2000	162738	–	162738	4332172	–	4332172	26620.53	–
2000–2001	171658	–	171658	4649017	–	4649017	27083.02	–
2001–2002	191796	–	191796	6097532	–	6097532	31791.76	–
2002–2003	178355	–	178355	5933174	–	5933174	33266.09	–
2003–2004	156061	–	156061	4916336	–	4916336	31502.66	–
2004–2005	152875	–	152875	4902387	–	4902387	32067.94	–
2005–2006	185888	–	185888	6709112	–	6709112	36092.23	–
2006–2007	159789	–	159789	5407235	–	5407235	33839.85	–
2007–2008	51040	–	51040	1713654	–	1713654	33574.73	–
2008–2009	56286	–	56286	2014909	–	2014909	35797.69	–
2009–2010	99606	–	99606	3866499	–	3866499	38817.93	–
2010–2011	109495	–	109495	4702825	–	4702825	42950.13	–
2011–2012	96350	–	96350	4069845	–	4069845	42240.22	–
2012–2013	82516	–	82516	3467395	–	3467395	42020.88	–
2013–2014	97101	–	97101	4730988	–	4730988	48722.34	–
2014–2015	105036	–	105036	5594240	–	5594240	53260.22	–
2015–2016	110204	–	110204	5965628	–	5965628	54132.59	–

Chapter 3
Performance Measurement and a Review of Literature

Abstract This chapter is divided into two parts. The first part introduces the theoretical framework of technical, allocative and their combinations' (economical) efficiency measurement based on input-oriented (cost minimization) or output-oriented (output maximization) measurements. This part addresses some of the drawbacks of cross-sectional data in technical efficiency measurement and addresses the advantages of using panel data instead. The second part of the chapter provides a detailed review of literature on measuring technical efficiency. This chapter makes three main contributions to literature. First, this book fills the gap between a methodological analysis of panel data stochastic frontier models and its empirical applications in the agricultural sector by reviewing existing stochastic panel data models in literature. Second, the book considers the effect of time in non-parametric models using a Window analysis. Time effect is usually ignored in non-parametric frontier models, especially in new partial frontier models (for example, Order-m and Order-α models). Third, the book not only compares different parametric and non-parametric models but it also chooses the models which are best suited to the datasets that are used based on different tests.

3.1 Introduction

In an economy, efficiency is a tool for optimal allocation of resources. This chapter uses an efficiency analysis to evaluate the possibility of increasing products by preserving available resources. In theoretical literature, this economic concept has been defined and measured in terms of technical, allocative and economic efficiency. This chapter first discusses the theoretical frameworks of efficiency before presenting a literature review. The literature review is done in two groups of national and international efficiency studies.

© Springer Nature Singapore Pte Ltd. 2018

M. Rashidghalam, *Measurement and Analysis of Performance of Industrial Crop Production: The Case of Iran's Cotton and Sugar Beet Production*, Perspectives on Development in the Middle East and North Africa (MENA) Region, https://doi.org/10.1007/978-981-13-0092-9_3

3.2   Performance Measurement

3.2.1   Definition of Efficiency

Efficiency is an economic concept which shows the performance of economic activities within an enterprise or an economic sector or a national (or regional) economy. Technical efficiency indicates the ability of a firm or decision making unit to: (1) produce the maximum products using a certain amount of inputs and (2) employ the minimum amount of inputs for producing a certain amount of output. It can also be defined as the ratio of the actual output to the optimal (maximum) output level at a certain level of inputs or in terms of the ratio of actual consumption or use of the production factors to optimum (minimum) consumption at a certain level of outputs (Hakimipour and Hojabrkiani 2008). In general, in empirical studies the minimum and maximum are based on the sample forming frontiers.

According to most existing empirical evidence a producer has not always been successful in solving his/her optimization problems and does not have full performance in terms of efficiency. Even having technical efficiency is not reason enough for a'producer to benefit from other dimensions of efficiency. A violation of the full efficiency assumption results in errors in conclusions and arguments. Considering this problem, instead of paying attention to conventional functions in analyzing producer behavior economic analysts pay attention to frontier functions. Hence, the implicit intents of econometrics have been formulated again (Hakimipour and Hojabrkiani 2008).

Recognizing an efficient production function is essential in technical efficiency measurement. Usually, the frontier is not revealed and it should be estimated using available data. Figure 3.1 presents different methods of technical efficiency measurement. Based on the different assumptions in the estimation of the production frontier, efficiency measurement is divided into parametric and non-parametric approaches.

Aigner et al. (1977) are pioneers of the parametric efficiency approaches. In these approaches, the production function is specified with a random error term and estimated by econometric methods to finally measure the efficiency and inefficiency of decision making units (DMUs). The three parametric approaches are the stochastic frontier approach (SFA), a thick frontier analysis (TFA) and a distribution-free analysis (DFA). Banker et al. (1984) are pioneers of non-parametric approaches for measuring technical efficiency utilizing linear programing. In this analytic approach, efficiency is measured using linear programing without specifying a functional form. This approach aims to find a frontier to overspread all DMUs in an industry and applies this frontier as a basis for measuring efficiency. The non-parametric approaches include full frontier and partial frontier analyses. Data envelopment analysis (DEA) and free disposal hull (FDH) are two classes of full frontier analysis. On the other hand, Order-m and Order-α efficiencies are two classes of a partial frontier analysis.

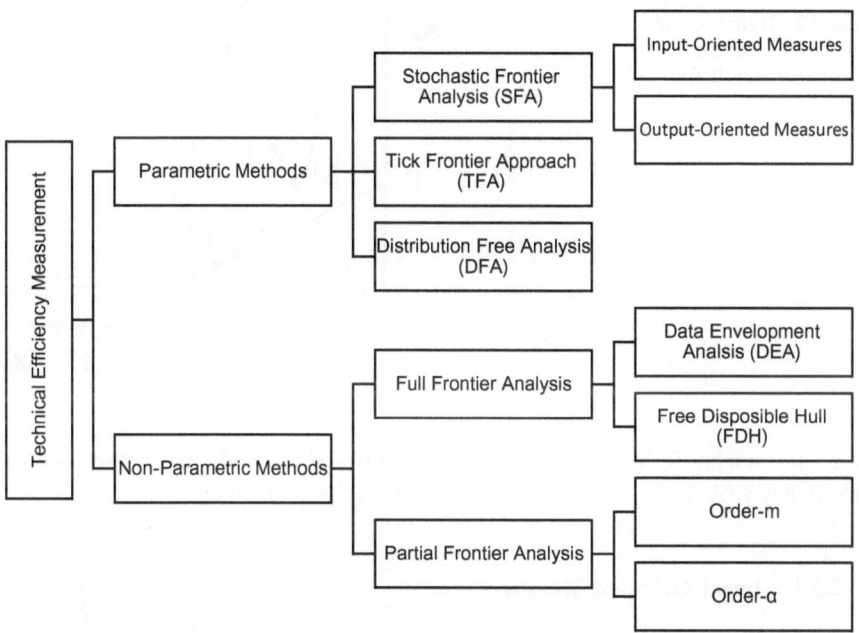

Fig. 3.1 The technical efficiency measurement framework. (Designed by Author from different sources)

3.2.2 *Parametric Methods*

Based on the deviations of an observation from the frontier, an estimated frontier can be deterministic or stochastic. Deterministic frontiers are based on regression and they associate all deviations with inefficiency. These frontiers are estimated by corrected ordinary least squares (COLS) or modified ordinary least squares (MOLS). The error term is divided into random and inefficiency components. In SFA, the error component is decomposed by parameterizing the distribution of the inefficiency term (Fried et al. 2008) which includes considering half-normal or exponential assumptions in the distribution of the inefficiency component (Coelli and Battese 1988; Emokaro and Ekunwe 2009). These are estimated using maximum likelihood (MLE). Besides, it is possible to use the conditional distribution of estimators to obtain the desired value of inefficiency for each DMU. The most important weakness of SFA is its parametric inherent (Chiami 2011). Before starting an analysis, the necessity of determining the proper functional form for the production function leads to biased results. Scholars also maintain that SAF compares each DMU with the average frontier, rather than with the best practice DMU. Moreover, according to Collier et al. (2011) technical efficiency cannot be analyzed in multidisciplinary terms without considering product data. In case of a multi-product, output variables are usually measured as total monetary value rather than physical units. Also, the

Fig. 3.2 Technical and
allocative efficiencies.
Source: Coelli (2008)

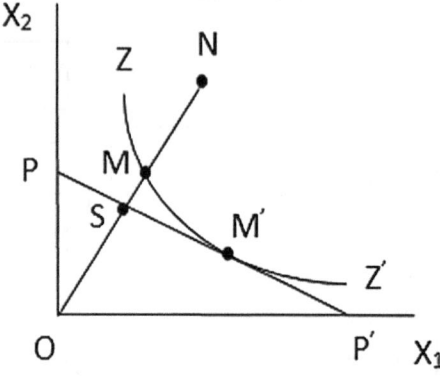

use of a stochastic frontier analysis in a multi-product mode may underestimate
inefficiency (Ajibefun 2008; Alene et al. 2006; Solís et al. 2009).

3.2.2.1 Input-Oriented Measures

Farrell (1957) considered a firm which used two inputs (x_1 and x_2) to produce one
output (y) under the CRS assumption. In Fig. 3.2, the set of efficient enterprises is
shown by ZZ' which indicates all the possible combinations of inputs that produce
a given level of output (Coelli 2008). Therefore, three regions can be defined in
this figure: (i) the upper points of ZZ' (inefficient firms), (ii) on the ZZ' (efficient
firms) and (iii) below the ZZ' (inaccessible). The technical inefficiency of firm (N) is
equal to MN, which indicates the amount to which all inputs can be reduced without
reducing production. Usually this value is represented by the OM/ON ratio. The
technical efficiency of firm (N) is measured as[1]:

$$TE_I = OM/ON \qquad (3.1)$$

Accordingly, by expanding the firm's distance from ZZ', the technical efficiency
tends to zero, otherwise it moves to 1 or full efficiency. If the input price ratio, that
is, the PP' line is known, then allocative efficiency (AE) can also be calculated.
Allocative efficiency shows the capability of an enterprise to use the optimum mix
of inputs according to the price. The allocative efficiency of firm (N) is:

$$AE_I = OS/OM \qquad (3.2)$$

---

[1]Input-oriented technical efficiency.

Fig. 3.3 Piecewise convex isoquant. Source: Coelli (2008)

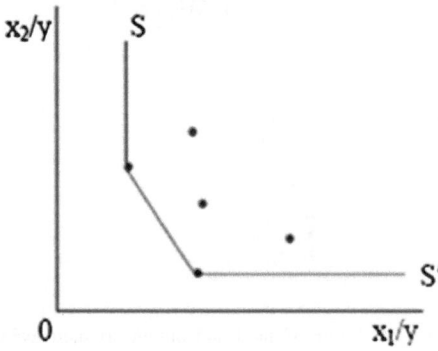

Economic efficiency (EE) is measured by multiplying the technical efficiency and allocative efficiency components. The economic efficiency of a firm operating at point N is:

$$TE_I \times AE_I = (OM/ON) \times (OS/OM) = (OS/ON) = EE_I \qquad (3.3)$$

The calculated efficiencies are based on the implicit assumption that the production function of a fully efficient firm is known. However, in practice this is not the case and the efficient isoquant should be estimated using sample data. Farrell presented two solutions for this: (i) constructing a piecewise-linear convex isoquant curve in which no observations should lie in the left or below part (Fig. 3.2); and (ii) estimating the parametric production function based on sample data (Fig. 3.3) in which no observations should be placed in the left or lower part (Coelli 2008).

3.2.2.2 Output-Oriented Measures

Technical efficiency measurement in terms of inputs helps answer the following questions: 'How much can inputs be reduced to make no difference to the quantity of output?' and 'How much is it possible to increase output without altering the quantity of inputs?' The answers to these questions is output-oriented measures which are opposed to input-oriented efficiency measures. The two approaches converge to the same optimal point or solution. The difference between these two measures is shown in Fig. 3.4. In Fig. 3.4a, there is decreasing returns to scale and production technology is shown with f(x). Considering inefficient firm P, its technical efficiency in terms of input and output measures are AB/AP and CP/CD respectively.

If we have CRS, measuring the efficiency will be the same by both the methods. Otherwise, the results will be different (Fare and Lovell 1978). The constant returns to scale is shown in Fig. 3.4b in which for any inefficient firm (P), $AB/$AP = CP/CD.

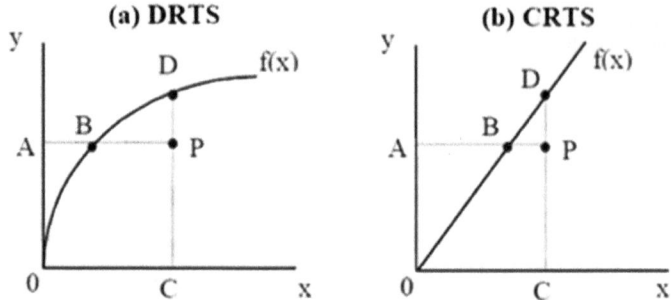

Fig. 3.4 Input-oriented and output-oriented technical efficiency measures and returns to scale. Source: Coelli (2008)

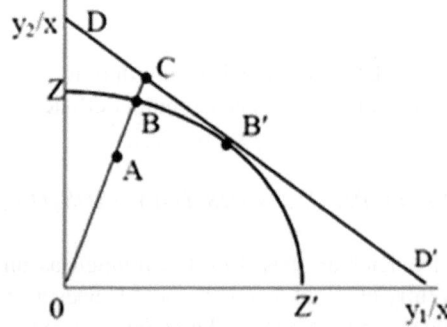

Fig. 3.5 Technical and allocative efficiencies from an output orientation. Source: Coelli (2008)

In case of two outputs (y_1, y_2) and one input (x_1), it is possible to show the technology by the two-dimensional production possibility frontier (PPF). As seen in Fig. 3.5, where ZZ' is PPF, A presents an inefficient firm. It should be noted that considering ZZ' as the upper limit of production possibility, every inefficient enterprise firm, such as A, lies below the curve.

In Fig. 3.5, the distance of AB indicates technical inefficiency. The distance shows the value by which output can be increased using extra inputs. The output oriented technical efficiency measurement is written as[2]:

$$TE_O = OA/OB \tag{3.4}$$

Iso-revenue line DD' can be drawn by price data and allocative efficiency is written as:

$$AE_O = OB/OC \tag{3.5}$$

---

[2]Output-oriented technical efficiency.

Also output-oriented economic efficiency (EE) can be obtained by:

$$EE_O = (OA/OC) = (OA/OB) \times (OB/OC) = TE_O \times AE_O \qquad (3.6)$$

Here it is essential to mention two important facts about the proposed methods (Coelli 2008). First, mentioned efficiency measures are obtained along a ray from the center to the production point. Therefore, the ratio of inputs (or products) is constant over that length. Any benefit of the radial efficiency measures is that it is unit-invariant, that is, the efficiency value does not change by changing the unit of measurement. In non-radial measurements, technical efficiency changes by reforming the units of measurement. Any change in the units of measurement leads to the selection of the 'nearest' point. This issue can be discussed in the evaluation of slacks in DEA. Second, the Farrell output and input-oriented measures are the same as distance functions of inputs and outputs introduced by Shepherd (1970).

3.2.3 Non-parametric Methods

Non-parametric approaches use mathematical programing methods to measure the relative efficiency of DMU. DEA and FDH are the most common methods of the non-parametric approach's efficiency measurement. A piece-wise frontier is formed based on the point that to produce a certain amount of product, the minimum amount of inputs should be used. It measures relative efficiency by comparing the true yield with the best yield. The DEA method applies general assumptions about monotonicity and convexity and generates a flexible frontier in which the functional form of the product can change among the MDUs. In the FDH method, the convexity of PPF is released and it has a step-wise frontier (Chiami 2011).

3.2.3.1 Data Envelopment Analysis (DEA)

DEA is a non-parametric mathematical programing method for estimating a frontier. This method evaluates the estimation of the frontier by a piecewise-linear convex hull method which was proposed by Farrelll (1957) and later examined by a number of researchers. Although Afriat (1972) and Boles (1996) suggested using mathematical programing methods for this purpose, this has not been considered much since the publication of Charnes, Cooper and Rhodes' (CCR) article (1978). CCR (1978) presented an input-oriented model with constant returns to scale (CRS) assumption. In later research, Banker, Charles and Cooper (BCC) (1984) proposed the variable return to scale (VRS) model (Coelli 2008).

Constant Return to Scale (CRS)

Technical efficiency of N firms with K inputs and M products is calculated as:

$$Max_{u,v} \left(u'y_i / v'x_i \right)$$
$$\text{st } u'y_j / v'x_j \leq 1, j = 1, 2, \ldots, N,$$
$$u, v \geq 0.$$
(3.7)

where u is a M × 1 vector and indicates the product's weight and v is a K × 1 vector containing weights of production factors. Besides, x is a K × N matrix of inputs and y is a M × N matrix of outputs with the aim of obtaining optimal values of u and v such that the weighted sum of the outputs is maximized to the weighted sum of the inputs (unit's rate of efficiency) only if the technical efficiency score is smaller or equal to unity.

This comes with one constraint as the u and v vectors are unknown. The only problem with this ratio is that it has an infinite number of solutions. To solve this problem, we can impose restriction $v'x_i = 1$:

$$Max_{\mu,v} \left(\mu'y_i \right),$$
$$\text{st } v'x_i = 1$$
$$\mu'y_j - v'x_j \leq 0, j = 1, 2, \ldots, N,$$
$$\mu, v \geq 0,$$
(3.8)

where μ and v are replaced with U and V. In linear programing, this notation is known as 'multiplier' and can be solved using the duality in linear programing:

$$Min_{\theta,\lambda}\theta,$$
$$\text{st } -y_i + Y\lambda \geq 0,$$
$$\theta x_i - X\lambda \geq 0,$$
$$\lambda \geq 0,$$
(3.9)

where λ is a N × 1 vector of constants and θ is a scalar.

The linear programing is solved N times, once for each DMU and a value of θ is obtained for each firm. Besides, the envelopment form has fewer constraints as compared to the multiplier form ($K + M < N + 1$), therefore, this from is the more preferred one (Coelli 2008).

Variable Returns to Scale (VRS)

The efficiency analysis of enterprises in CRS and VRS forms are considered as long-term and short-term status. The CCR model with the assumption of CRS presents technical efficiency which contains pure technical efficiency (management efficiency) and efficiency resulting from the scale saving of an enterprise. The CCR model was developed by Banker, Charnes and Cooper in 1984 and VRS was added to the model by adding a convexity constraint ($N1'\lambda = 1$) to previous linear programing:

$$Min_{\theta\lambda}\theta,$$
$$st -y_i + Y\lambda \geq 0,$$
$$\theta x_i - X\lambda \geq 0, \tag{3.10}$$
$$N1'\lambda = 1,$$
$$\lambda \geq 0,$$

3.2.3.2 Free Disposal Hull (FDH)

The FDH model was first used by Deprins and Tulkens (1984) in technical efficiency literature. By using the FDH technique one can rank the efficiency of DMUs by comparing each DMU's performance with a PPF. FDH is a non-parametric method and the full frontier and no convexity condition for production-possibility set is imposed. Besides, in measuring the efficiency of each production possibility frontier, only that unit is compared with another one. This means that in this model the reference unit for each inefficient unit is one of the sample units (Afonso and Aubyn 2005).

Consider Fig. 3.6 which shows the efficiency of four hypothetical firms. A firm operating at point D is an inefficient one compared to that operating at point C as it produces less output using more inputs. In FDH, firms A, B and C are efficient firms (Afonso et al. 2003).

3.2.3.3 Order-m and Order-α

Proponents and opponents of parametric and non-parametric methods argue about important issues and suggest new alternative approaches. The most important drawback of parametric methods is that they impose limiting assumptions on the functional form and the distribution of random error and inefficiency terms. Besides, in parametric approaches the amount of inputs is the center of importance as explanatory variables which are likely to be endogenous. Also, due to complexity in the modeling and estimation of these techniques only the single product function is considered. On the other hand, non-parametric methods are deterministic and are sensitive to measurement errors and outliers. Econometricians criticize these methods on these grounds (Tauchmann 2011). Individual heterogeneity, individual inefficiency and random error terms are all confounded biasing efficiency.

As mentioned earlier, one of the most important drawbacks of non-parametric models is their sensitivity to outliers and measurement errors. This was solved by the 'partial frontier', that is, Order-m (Cazals et al. 2002) and Order-α (Aragon et al. 2005). These two methods generalize the FDH model using the super-efficient observational placement along the production-possibility frontier. Therefore, the partial frontier models are less sensitive to outliers and measurement errors compared to DEA or FDH approaches (Tauchmann 2011).

In full frontier models in which all DMUs are enveloped by the production-possibility frontier we have: $\theta_k^{inp} \in (0, 1]$ and $\theta_k^{out} \in [0, \infty)$. Therefore, the efficiency of efficient DMUs equals 1 and any downward (in input-oriented) and upward (in output-oriented) deviations from unity indicates inefficiency. Conversely, in partial frontier methods, inefficiencies can be higher (input-oriented) or lower (output-oriented) than 1 (Tauchmann 2011).

Figures 3.7 and 3.8 show non-parametric frontier approaches discussed in the previous section. Figure 3.7 shows the production-possibility frontier for 40 DMUs. For 36 DMUs, data was generated using the Cobb-Douglas technology. Besides, for four DMUs, the inputs were not compatible with this technology and according to the true frontier input consumption was very low. These DMUs represent outliers.

Figure 3.7 shows input-oriented efficiency. For example, consider a DMU that operates at point A in which the true efficiency is equal to $\overline{OA^*}/\overline{OA}$. However, since the true frontier is unknown an estimated value is required.

Figure 3.8 shows estimated frontiers using DEA, FDH, Order-m and Order-α. In DEA and FDH models, outliers span the estimated frontiers and lead to other DMUs being highly inefficient. In other words, normal observations do not affect the estimated DEA and FDH frontiers. In contrast, in Order-α (=95α) and Order-m (m = 12) patterns, outliers fall outside the estimated frontier. Therefore, Order-α and Order-m use the information in normal observations to estimate the frontier which ultimately leads firms to be compared with a more appropriate benchmark (Tauchmann 2011).

Fig. 3.6 Comparison of CRS, VRS and FDH. Source: Afonso et al. (2003)

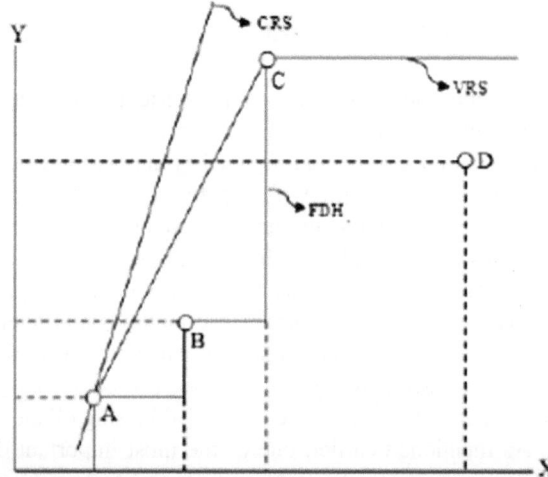

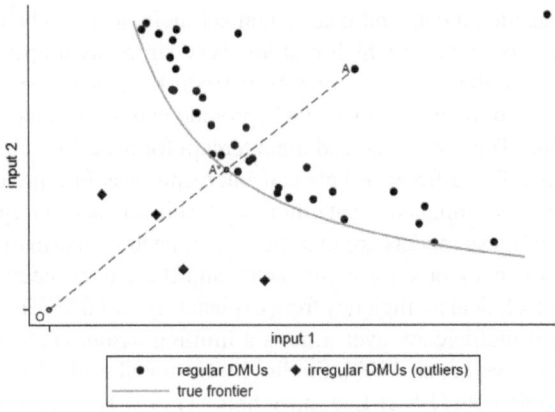

Fig. 3.7 Scatter plot of input use and true production possibility frontier (isoquant). Source: Tauch-mann (2011)

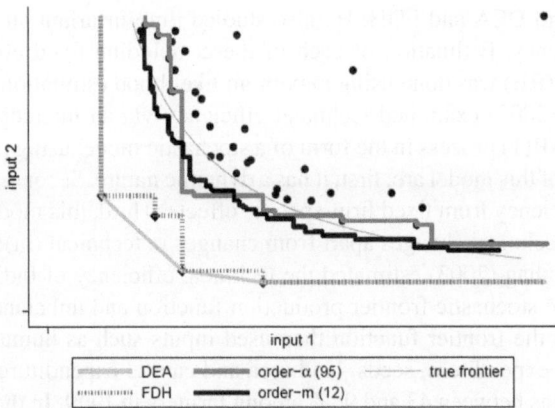

Fig. 3.8 Non-parametrically estimated production possibility frontiers (isoquant). Source: Tauch-mann (2011)

3.3 Literature Review

Studying and reviewing literature helps in getting to know the inadequacies and results of previous studies. Besides, by providing detailed information about a variety of methodologies the studies open up avenues for choosing the most complete and efficient ones. Therefore, to attain the goals of this book, reviewing the research background is of central importance. This section presents a summary of studies on the evaluation of technical efficiency using parametric and non-parametric methods.

Primary literature on frontier production functions (for example, Aigner et al. 1977; Greene 1980, 1990; Jundro et al. Meeusen and van den Broeck 1977; Richmond 1974; Schmidt and Lovell 1979; Schmidt and Sickles 1984; Stevenson 1980, 1982) considers cross-sectional data and takes technical inefficiency to be independent of the random error components which requires very strong assumptions about their distribution. Later studies (Battese and Coelli 1988; Kumbhakar 1987; Hallam and Machado 1995; Schmidt and Sickles 1984) provide extended models as panel data was now available. They only changed the intercept for firms in their time-invariant efficiency models. The difference between intercepts was interpreted as different levels of efficiency assuming constant efficiency levels for each enterprise over time. The advantages of these models are that there is no need to distribution assumption on technical inefficiency or error components and there is no need to assume the independence of technical inefficiency from explanatory variables (inputs). However, constant technical inefficiency over time is a limiting assumption and may not be true in reality (Cornwell et al. 1990). Studies by Cornwell et al. (1990), Kumbhakar (1990), Battese and Coelli (1992), Lee and Schmidt (1993), Battese and Coelli (1995), Ahn et al. (2001, 2007) were the first ones that evaluated time-varying technical inefficiency.

Henderson (2003) measured technical efficiency using panel data, the SA production function and DEA and FDH. He also studied time-invariant and time-varying technical efficiency. Estimation of each of these including fixed effects (FE) and random effects (RE) was done using maximum likelihood estimation (MLE).

Desli et al. (2003) examined technical efficiency via an intercept that evolved over time as a AR(1) process in the form of a stochastic model using panel data. The characteristics of this model are: first, it has a dynamic nature. Second, it can separate technical inefficiency from fixed firm-specific effects. Third, this model can be used for estimating technical changes apart from changes in technical (in)efficiencies.

Goyal and Suhag (2003) estimated the technical efficiency of Indian wheat producers using the stochastic frontier production function and unbalanced panel data for 1996-99. In the frontier function they used inputs such as human labor, fertilizers, irrigation expenditure, seeds, land area and capital expenditure. Their results showed variations between 43 and 95% among farmers in 1999. In the study period, the average technical efficiency had a decreasing trend—from 91% in 1996 to 90% in 1999.

Ahn et al. (2007) proposed a time-varying SFA model and assumed that the technical efficiency of firms was time-varying. They called the studied model 'multiple time-varying individual effects' which was more general. Also, this model led to the estimation of the technical efficiency of a firm that was firm-specific and time-specific without assuming the same pattern of temporary changes in efficiency for all enterprises. Therefore, it allowed a temporary pattern of inefficiency changes over firms. The researchers used the proposed model for rice producers in Indonesia. Their results indicated that rice producers not only had different levels of performance, but also different efficiency temporal patterns.

Kamruzzaman and Islam (2008) evaluated technical efficiency and its determinants in wheat production in Bangladesh in 2004 using the Cobb-Douglas stochastic frontier production function. Based on their research findings, the technical efficiency range of areas under wheat cultivation varied from 40 to 99% and the technical efficiency level was on average 70.33%.

Lambarraa (2012) evaluated dynamic technical efficiency in the Spanish horticultural sector for both greenhouse and non-greenhouse products. According to their results, the static efficiency level for greenhouse farms was 12.0% higher than non-greenhouse farms. The estimated level of static efficiency suggested that in the studied period greenhouse and non-greenhouse fields had 15.0 and 11.0% increase and decrease in efficiency respectively. This difference can be explained by a proper control over the use of different inputs and the use of appropriate technologies for improving efficiency levels in greenhouse fields over the years. Experimental results also indicated a big difference in the dynamic and static technical efficiency for both the samples.

Kumbhakar (2014) examined six different models based on various assumptions and specifications of heterogeneity, heteroscedasticity and technical inefficiency. He estimated all these models using information on outputs and inputs for Norwegian grain producers for the period 2004–08. He also presented a new model which extracted firm effects from persistent and residual technical inefficiency. His findings showed that efficiency scores were quite sensitive to the type of interpretation and inefficiency modeling.

Mohammed and Saghaian (2014) evaluated the technical efficiency of rice producing provinces in South Korea using the stochastic frontier production function. According to their results, there was no significant difference in technical efficiency between different provinces and this varied from 79.0% in Jeollanam-do province in 2012 to 99.0% in Chungcheongbuk-do province in 1993. Excessive use of seeds and machinery in rice production was the reason for the inefficiency of this product. According to their results, it is possible to increase production efficiency by replacing the family workforce with a recruited (hired) labor force. The effect of location on technical efficiency was also significant.

Although DEA has been widely used over the past two decades, the Window DEA analysis has been relatively rare in literature. Gabdo et al. (2014) used two full frontiers (DEA and FDH) and two partial frontiers (Order-α and Order-m) techniques to do a comparative estimation of technical efficiency. According to their results, although the estimators were different in the assumptions, the estimation of technical efficiency obtained from the four completely different estimators was very similar.

Pjevcevic et al. (2012) analyzed technical efficiency and its determinants in the harbors of Serbia using Window DEA. They also examined the possibility of changing the Harbor's efficiency over time using this method. Yang and Chang (2009) calculated the technical efficiency of Taiwan telecom companies using Window DEA during 2001–05. Carbone (2000) used the Window DEA analysis to identify the efficiency of factories over time. Sueyoshi and Aoki (2001) studied the performance of the Japanese postal services by combining DEA Window and Malmquist Productivity Index during 1983–97. Ross and Droge (2002) used Window DEA to evaluate and identify the performance of broadcast centers over four years. Asmild et al. (2004) integrated Window DEA with the Malmquist Productivity Index and examined productivity variations in the Canadian banking industry during 1981–2000.

Řepková (2014) studied the efficiency of commercial banks in the Czech Republic using the Window DEA analysis over the period 2003–12. The author used Window

DEA based on the input-oriented model. The results indicated that the average efficiency under CRS and VRS was 70–78% and 84–89% respectively.

Song et al. (2016) studied productivity changes, technical efficiency and agricultural production technology in 13 provinces of China during 1999–2008. They used the three-step Malmquist Productivity Index approach. Their results showed that total productivity increased by 1.6% in Chinese agricultural products during the period. On the other hand, after applying supportive policies for the purchase price of cereals during the 1990s, total factor productivity of Chinese crop production began declining.

We now focus on a review of literature on technical efficiency measurement in Iran.

Yadollahi (2003) evaluated the efficiency and productivity of Iranian Industrial Factory using parametric and non-parametric methods. The results showed that the Iranian Industrial Factory showed no improvements in terms of technical efficiency and productivity.

Azar and Gholamrezai (2006) studied human development in different Iranian provinces using DEA. They also measured the efficiency of the provinces in using infrastructural resources to produce human development indicators in 2002. Their results indicated that deprived provinces were more efficient due to limited resources. They also suggested that more attention should be paid to deprived provinces in planning better allocation of resources.

Using SFA Hakimipour and Hojabrkiani (2008) estimated and compared the technical efficiency of the industrial sector (factories) in different provinces of Iran during 1991–2004. They also analyzed determinants of inefficiency and its differences in the provinces. Their results showed that the average provincial efficiency of the mentioned industries was 37.4%, which was a significant yield. The highest and the lowest average efficiency were in Khuzestan province (76%) and Sistan and Baluchestan provinces (19.9%) respectively.

Mojarad et al. (2011) calculated the technical efficiency of poultry units in Sistan and Baluchestan provinces in Iran using the stochastic non-parametric approach. They collected information on 41 active poultry units and estimated their technical efficiency by a combination of SFA and DEA methods. Their results showed that most of the poultry units were inefficient in terms of technical efficiency and the average efficiency of units was 94.0%. In addition, 48, 43 and 7% of the studied units had constant returns to scale, increasing returns to scale and decreasing returns to scale respectively. The results confirmed that increasing inputs' productivity and applying supportive policies in input and output markets will increase production and improve technical efficiency.

Zeranejad and Yousefi (2009) measured technical efficiency of wheat production in different provinces in Iran during 1999–2004 using a parametric approach. Their results showed an average technical efficiency of about 57% during the studied period. According to this study, Gilan and Bushehr provinces with 81 and 26% technical efficiency were the most and least efficient provinces in the country respectively. The results of the non-parametric model indicated that in the same period the average technical efficiency was about 84%. Sistan and Baluchestan,

Kohgiluyeh and Boyer-Ahmad, Gilan and Mazandaran with 100% and Yazd with 57% efficiency had the highest and lowest technical efficiency respectively.

Amadeh et al. (2009) estimated the industrial sector's technical efficiency in Iran's provinces using DEA and then ranked the provinces according to the results. Their results showed that Bushehr, Khuzestan, Hormozghan and Kerman had the highest technical efficiency, followed by Tehran, Isfahan, Markazi and East Azerbaijan. Based on this study, the average technical efficiency of the industrial sector in 28 provinces was about 62.7% during 1996–2004.

Yaghubi et al. (2010) studied the efficiency of cooperative and non-cooperative shrimp breeding units in the Chahbahar county in the Sistan and Baluchestan province using DEA and FDH methods in 2009. The results showed that only 12 and 16% of the enterprises were efficient based on the DEA and FDH models. Average technical efficiency based on these models was 85 and 87% respectively.

Abtahi and Islami (2010) evaluated the technical efficiency of 25 rainfed wheat producing provinces of Iran using SFA. Their results showed that technical efficiency depended on time. Average technical efficiency was about 50% which decreased during the study period. It should be noted that there are significant differences between technical efficiency in parts of Iran.

Sokhanvar et al. (2011) evaluated the efficiency process of power distribution companies using Window DEA and studied the structural and environmental determinants of efficiency. The companies were divided into two groups of low (group 1) and high circuit density (group 2). According to their results, the mean efficiency of the two groups was considerably upward and downward in relation to the transboundary under both the assumptions of constant and variables' returns to the scale respectively. Power distribution centers in Shiraz, Golestan and Mazandaran (group 2) were inefficient.

Zeranejad et al. (2012) studied technical efficiency and its determinants in the Iranian Industrial Factory using SFA. Their results showed that average technical efficiency was 55% during the study period. Active industries in the field of copper and refined petroleum products had higher average technical efficiency compared to the other sectors. In contrast, active industries in brick manufacturing and cereals and grains processing industries had the lowest technical efficiency. The average efficiency of most industries ranged from 50 to 60%.

Babai et al. (2012) measured technical efficiencies of greenhouses in Zabol province of Iran using DEA in 2009. Their results showed that average efficiency varied from 3.3 to 90.8%. The highest efficiency level was 100% and the lowest efficiency was 84.6% which indicated that by increasing farmers' technical efficiency, for example, through training, it was possible to increase production and reduce costs without significant changes in the levels of technology and resources.

Karami et al. (2012) measured the technical efficiency of 44 aquaculture, broiler and dairy cattle producing units in 2009. According to their results, broiler producing units' efficiency in Gachsaran, Dena, Bahmai and Kohgiluyeh provinces were in the first to third-grade rank. In aquaculture production units, Dena province ranked first and Boyer-Ahmad had the second rank. Besides, the average efficiency of dairy cattle production units in Bahmai, Boyer-Ahmad, Gachsarab, Dena and Kahgiluyeh provinces had the first to the fifth rank respectively. The average efficiency of the production units was about 85%. The results also indicated the differences between technical efficiencies in different provinces.

Durandish et al. (2013) evaluated the effective factors in barberry production and measured technical efficiency of barberry producers in South Khorasan in Iran. They used the Cobb-Douglas production function to study the effects of these effective factors on barberry production. Their results showed that barberry producers' average technical efficiency was about 81% and the highest and lowest technical efficiencies were 99 and 34% respectively. Experience, main job and the number of family members working had a significant effect on technical efficiency. Therefore, increasing irrigation and farmers' experience are two important factors in barberry production. In this regard they suggested the use of new irrigation techniques and holding training classes to enhance barberry farmers' experience.

Motafakerazad et al. (2014) measured technical efficiency and its determinants in Iran's thermal power stations using the StoNED method during 1999–2011. Their results showed that size and operating rate had positive while age had a negative effect on the technical efficiency of thermal power stations. Besides, gas-based power plants had higher efficiency. Their findings indicate that the restructuring of the electricity market in 2003 had a positive effect on the technical efficiency of Iran's thermal power plants.

Hoseinpour et al. (2014) evaluated the technical efficiency of rosewater in Kashan in 2004. The production function showed that 22% of the producers overused the flower input and 28% of the producers produced it in the first region of production. The average technical efficiency of producers was 97%, ranging from a minimum of 67% to a maximum of 99%. Age, experience, acquisition of new sciences and secondary incomes were determinants of efficiency.

Behroz and Emami Meybody (2014) measured technical, allocative and economic efficiencies and also the productivity rate of irrigated watermelon producers in 12 provinces using the non-parametric DEA and the Malmquist Productivity Index. Their results showed that average technical, allocative and economic efficiencies of irrigated watermelon producers during 2005–10 was about 79.4, 75.9 and 61.5% respectively.

Ardabili-Mianji and Barimnezhad (2016) evaluated the efficiency of 25 branches of agricultural banks in Alborz province during 2012–13 using two basic DEA models (Output-oriented BBC and CCR models). According to the results of this research, the average technical efficiency of agricultural banks in this province in terms of constant and variable returns to scale in 2012 was about 84.2 and 87.7% respectively. In 2013 this was about 92.8 and 93.7% respectively. Also, the average scale efficiency was about 95.9 and 98.9% in 2012 and 2013 respectively.

A review of the studies done in Iran indicates that most of the studies used cross-sectional data. Only a few used panel data and that too simple and fundamental models of Pitt and Lee (1981), Cornwell et al. (1990) and Battes and Coelli (1992). Moreover, they by and large used models with two-component error elements. The research presented in this book with a four-component error element and the type of technique that it uses is novel. Besides only the Window-DEA method has been used in literature to calculate firm-specific efficiency and there is limited research on Window-FDH, Window Order-m and Order-α models. Therefore, it is necessary to conduct a comprehensive study on the production status of industrial crops in Iran using these new modeling and estimation techniques. This research is unique in terms of the comprehensiveness of the models used in current literature on the country. It is also necessary that the provinces in the country be ranked in terms of technical efficiency.

Chapter 4
Parametric and Non-parametric Models for Efficiency Measurement

Abstract The first part of this chapter outlines 12 parametric panel data models classified into four groups in terms of the assumptions made on the temporal behavior of inefficiency. The models are grouped by their underlying assumptions: models assuming the inefficiency effects to be time-invariant and individual-specific; allowing inefficiency to be individual-specific but time-varying; models separating inefficiency effects from unobserved individual effects; and models separating persistent inefficiency and time-varying inefficiency from unobservable individual effects. In general, all performance measurement methods are expected to generate individual-specific effects. The chapter sheds light on whether particular panel data stochastic frontier models are better suited to different datasets.The second part of this chapter includes four non-parametric models. Two of these models are full frontier models (DEA and FDH) and the other two are partial frontier models (Order-m and Order-α).

4.1 Introduction

The frontier function methodology has been given particular attention for measuring and comparing the performance of decision making units in comparing countries, within a geographic location, a service sector, an industry or firm and plant levels. Extensive research in this field has resulted in the rapid development of econometric techniques concerning specifications, estimations and testing issues. These techniques have been developed rapidly and implemented in a large number of areas using cross-sectional and panel data. Some of the problems related to distributional assumptions encountered in the cross-section approach are avoided in panel data models. Panels also give a large number of data points and have the advantage of separating individual and time-specific effects from the combined effect (Heshmati et al. 1995). According to Schmidt and Sickles (1984), another advantage of panel data is that with a time-invariant inefficiency assumption one can estimate inefficiency consistently without distributional assumptions. Time-invariant inefficiency

---

Some parametric models of this chapter are discussed in Rashidghalam et al. (2016).

© Springer Nature Singapore Pte Ltd. 2018   41
M. Rashidghalam, *Measurement and Analysis of Performance of Industrial Crop Production:*
The Case of Iran's Cotton and Sugar Beet Production, Perspectives on Development in the Middle
East and North Africa (MENA) Region, https://doi.org/10.1007/978-981-13-0092-9_4

assumption is quite strong, although the model is relatively simple to estimate if efficiency is specified as a fixed parameter instead of as a random variable (Battese and Coelli 1988; Kumbhakar 1987; Pitt and Lee 1981). The other extreme is assuming that both inefficiency and noise terms are independently and identically distributed (i.i.d.). This assumption makes the panel nature of the data irrelevant. Some models fall between these extreme (e.g. Karagiannis and Tzouvelekas 2009; Kumbhakar et al. 2014).

In non-parametric methodology Charnes et al. (1978) were the first to develop the data envelopment analysis (DEA) Model, which develops an empirical frontier using observed production and measures technical efficiency as the distance of each decision making unit (DMU) from the frontier. DEA has the advantage of handling multiple outputs and inputs without price information and functional forms (Ruggiero 2007). Subsequently, Banker et al. (1984) proposed a flexible model in which it is assumed that returns to scale is a variable (VRS). Free Disposal Hull (FDH) model is the other model which is used extensively in non-parametric technical efficiency measurement. FDH was first formulated by Deprins et al. (1984) and then developed by Tulkens (1993). The advantage of this model over DEA is that it is based on the principle of weak dominance and departs from the convexity assumption inherent in the DEA model. This model also assigns an already existing DMU an efficient reference point, which makes the achievement of goals more credible. However, it also marks more DMUs efficient.[1] The DEA and FDH models are considered as full frontier models. These models assume that all observations belong to an attainable set and are based on the envelopment approach. Comparison of parametric and non-parametric approaches shows that non-parametric methods have some advantages over parametric methods. Meanwhile, non-parametric methods have some limitations and have been criticized by econometricians for lacking a well-defined data generating process, being deterministic and being highly sensitive to extreme data and measurement (Carvalho and Marques 2014). By introducing partial frontier approaches, namely Order-m and Order-α, some of these objections are addressed (Cazals et al. 2002; Aragon et al. 2005). These two techniques generalize the FDH model by allowing for super-efficient decision making units to be located beyond the estimated production possibility frontier. Hence, the estimated frontiers will not be shaped by some outliers which might represent measurement errors. Therefore, partial frontier models are less vulnerable to outliers as compared to DEA or FDH (Tauchmann 2011).

Figure 4.1 shows the summary of the models used to analyze the technical efficiency of cotton and sugar beet producing provinces in Iran.

---

[1] See Cooper et al. (2007) for a comprehensive discussion on DEA and FDH.

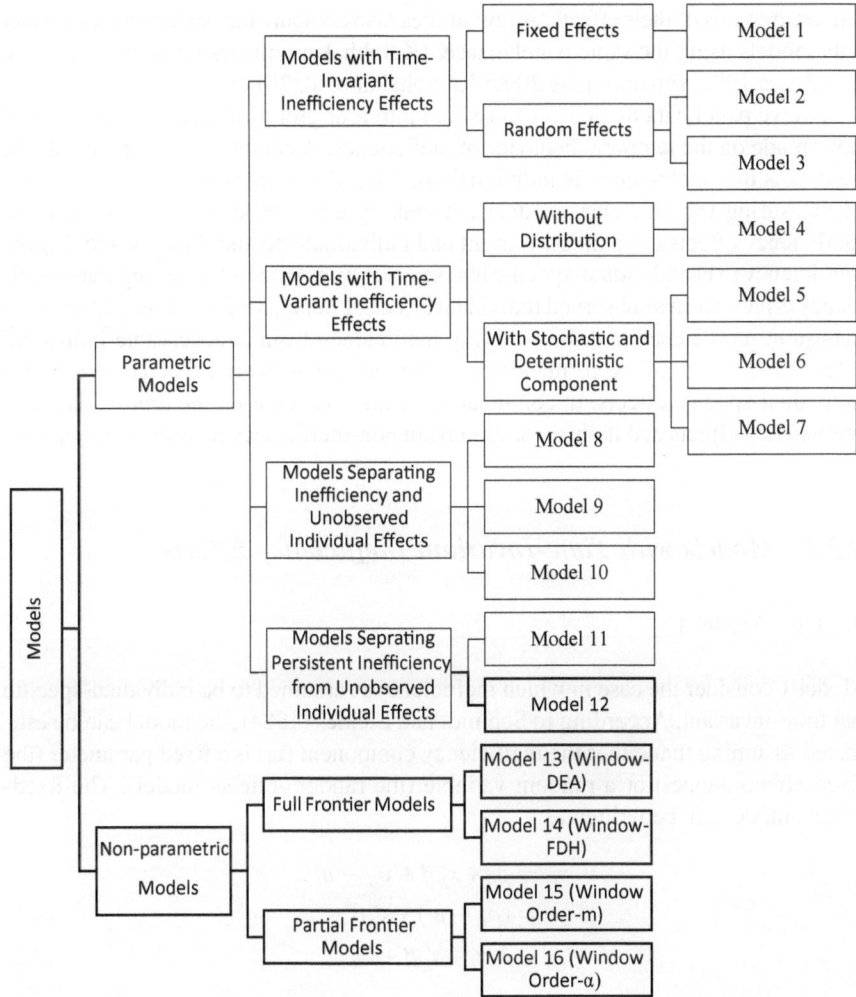

Fig. 4.1 Summary of the Models

4.2   Parametric Models

Parametric approach involves modeling the production process using econometric methods and taking into account the initial assumption of the functional form of the production and distribution of the inefficiency component. Cobb-Douglas and Translog functional forms are the most commonly used forms of the production function (Kumbhakar and Lovell 2000). Literature has developed enough to account for time-variance, heteroscedastic, persistent technical efficiency effects, separation of inefficiency and individual effects and identification of determinants of inefficiency and estimations of their effects. A few studies also compare the performance of panel data models using the same panel dataset (Kumbhakar and Heshmati 1995; Battese and Broca 1997; Emvalomatis 2009; Kumbhakar et al. 2014).

Twelve panel data models are classified into four groups in terms of the assumptions made on the temporal behavior of inefficiency. A common issue among all the models is that inefficiency is individual-specific. This is consistent with the notion of measuring the efficiency of decision-making units. Models 1 to 3 assume the inefficiency effects to be time-invariant and individual-specific. Models 4 to 7 allow inefficiency to be individual-specific but time-varying. Models 8 to 10 separate inefficiency effects from unobserved individual effects. Finally, Models 11 and 12 separate persistent inefficiency and time-varying inefficiency from unobservable individual effects. In general, all performance measurement methods are expected to generate individual-specific effects. In continuation of this, we focus on the time-variance of inefficiency effects and their separation from non-inefficiency heterogeneity effects.

4.2.1   Models with Time-Invariant Inefficiency Effects

4.2.1.1   Model 1

Model 1 consider the case in which inefficiency is assumed to be individual-specific but time-invariant. According to Schmidt and Sickles (1984), the model can be estimated assuming that either the inefficiency component (u_i) is a fixed parameter (the fixed-effects model) or a random variable (the random-effects model). The fixed-effects model can be written as:

$$
\begin{aligned}
y_{it} &= \beta_0 + x'_{it}\beta + \upsilon_{it} - u_i \\
&= (\beta_0 - u_i) + x'_{it}\beta + \upsilon_{it} \qquad (4.1) \\
&= \alpha_i + x'_{it}\beta + \upsilon_{it} \qquad (4.2)
\end{aligned}
$$

where $i = 1, \ldots, n$ denotes individual production units and $t = 1, \ldots, T$ denotes time periods. In the stochastic frontier model above, y_{it} is the logarithm of output for a decision making unit i at time t; x_{it} is the vector of inputs (in logarithm); β_0 is a common intercept β is the associated vector of technology parameters to be estimated; υ_{it} is a random two-sided noise term that can increase or decrease output (ceteris paribus); and $u_i \geq 0$ is the non-negative one-sided inefficiency term. Model (4.2) looks similar to a standard fixed-effects (FE) panel data model.

Once estimates of $\hat{\alpha}_i$ are available, the following transformation is used to obtain an estimated value of the time invariant individual specific inefficiency effects $\hat{u}_i$:

$$
\hat{u}_i = \max_i \{\hat{\alpha}_i\} - \hat{\alpha}_i \geq 0, \; i = 1, \ldots, N \qquad (4.3)
$$

The Eq. (4.3) implicitly assumes that the most efficient unit in the sample is 100 per cent efficient. Firm-specific efficiency measurement can be obtained as:

$$T\hat{E} = \exp(-\hat{u}_i), \ i = 1, \ldots, N \tag{4.4}$$

The most important weakness of this model is its strong assumption on time-invariant inefficiency and also its inability to separate inefficiency and individual producers' heterogeneity.

4.2.1.2 Model 2

In (4.2) it is also possible to assume that α_i is random and uncorrelated with X_{it} (random-effect model). If the assumption of no correlation is correct then the random-effect (RE) model provides more efficient estimates. The random-effect model can be estimated by two different methods: (i) Using generalized least squares (GLS) technique which is commonly used for a standard RE panel data models. In using this technique as the FE estimator, the RE is modified and re-interpreted to obtain inefficiency estimates, and (ii) Using maximum likelihood (ML) method (Model 3 in this chapter).

Assuming u_i as a random variable and letting $E(u_i) = \mu$ and $u_i^* = u_i - \mu$. The model in (4.1) can be rewritten as:

$$\begin{aligned}
y_{it} &= \beta_0 + x'_{it}\beta + v_{it} - u_i \\
&= (\beta_0 - \mu) + x'_{it}\beta + v_{it} - u_i^* \\
&= \alpha^* + x'_{it}\beta + v_{it} - u_i^*, \quad \alpha^* = \beta_0 - \mu
\end{aligned} \tag{4.5}$$

This model has some advantages over Model 1. Model 2 allows for testing the assumption of fixed or random inefficiency and also it provides a possibility of estimating the model efficiently.

4.2.1.3 Model 3

Imposing some distributional assumptions on the random components of the model and estimating the parameters by the maximum likelihood estimation (MLE) method in an alternative method to the GLS (Pitt and Lee 1981).

For MLE, the model is written as:

$$\begin{aligned}
y_{it} &= f(x_{it}, \beta) + \epsilon_{it}, \\
\epsilon_{it} &= v_{it} - u_i, \\
v_{it} &\sim N(0, \sigma_v^2), \\
u_i &\sim N^+(0, \sigma_u^2).
\end{aligned} \tag{4.6}$$

The MLE method generates higher efficiency in estimation through the iteration procedure. It is at the cost of strong assumptions of half-normality inefficiency term (u) and normality of the random error term (v) component.

4.2.2 Models with Time-Variant Inefficiency Effects

4.2.2.1 Model 4

The models of Group 1 assume technical inefficiency to be time-invariant and individual-specific. Meaning that inefficiency levels might be different for different individuals, but they do not change over time. Therefore, these models suggest that an inefficient DMU never learns or is not able to reduce the level of its inefficiency over time. This might happen in some situations where inefficiency is associated with managerial abilities and there is no change in the management for any of the firms during the period of the study. Also inefficiency can be persistent if the time period of the panel is particularly short. When market competition is taken into account, this at times is unrealistic. In this regard, in order to accommodate the notion of productivity and improvements in efficiency we need to consider models that allow inefficiency to change over time. Hence, Models of Group 2 are introduced in which the inefficiency effects are time varying.

Recall the model in (4.2), where α_i representing a mixture term of inefficiency and individual effects is time-invariant (Schmidt-Sickles 1984). To make it time-varying, Cornwell et al. (1990) suggest replacing α_i by α_{it}. The model is known as CSS model:

$$\alpha_{it} = \alpha_{0i} + \alpha_{1i}t + \alpha_{2i}t^2. \tag{4.7}$$

Note that the parameters α_{0i}, α_{1i} and α_{2i} are individual-specific and t is the time trend variable. This model is represented as:

$$\begin{aligned} y_{it} &= \alpha_{0i} + x'_{it}\beta + \upsilon'_{it}, \\ \upsilon'_{it} &= \upsilon'_{it} + \alpha_{1i}t + \alpha_{2i}t^2, \end{aligned} \tag{4.8}$$

The form of this model looks like a standard panel data model. As Schmidt and Sickles's (1984) model, one can apply a whithin estimator in (4.8) in order to obtain consistent estimates of β', and the estimated residuals of the model ($\hat{\varepsilon}_{it} = y_{it} - x'_{it}\beta'$). A disadvantage of this model is that time-variant is a function of a time trend and therefore it is unable to capture possible (non-trend) fluctuations in inefficiency over longer periods.

4.2.2.2   Model 5

In Model 5, following generic formulation is used in order to discuss the various components in a unifying network:

$$y_{it} = f(x_{it}, \beta) + \epsilon_{it},$$
$$\epsilon_{it} = v_{it} - u_{it},$$
$$u_{it} = G(t)u_i, \qquad\qquad G(t) > 0$$
$$v_{it} \sim N(0, \sigma_v^2),$$
$$u_i \sim N^+(\mu, \sigma_u^2). \qquad\qquad\qquad (4.9)$$

where $G(t)$ is a function of time (t). In this model, inefficiency (u_{it}) changes over time and across individuals. u_{it} is composed of two distinct components: the non-stochastic time component ($G(t)$) and a stochastic individual component (u_i). It is the stochastic component, u_i, that utilizes the panel structure of the data in this model. The u_i component is individual-specific and the $G(t)$ component is time-varying and is common for all the individuals. Given $u_i \geq 0$ and $u_{it} \geq 0$ is ensured by having a non-negative $G(t)$. Here we consider some specific forms of $G(t)$ which are used in the literature (Models 5, 6 and 7 of this chapter). First cinsider Kumbhakar's (1990) model which assumes $G(t)$ as:

$$G(t) = \left[1 + \exp(\gamma_1 t + \gamma_2 t^2)\right]^{-1} \qquad\qquad (4.10)$$

where $G(t)$ can be monotonically increasing (decreasing) or concave (convex) according to the signs and magnitudes of γ_1 and γ_2. As in Model 4, inefficiency changes in Model 5 are time driven and a non-linear exponential function of time. However, the trend pattern is similar for all individuals and the differences in performance among individuals are due to inefficiency term (u_i). The random and non-linear nature of this model requires iterative estimation using maximum likelihood estimation method.

4.2.2.3   Model 6

An alternative formulation for $G(t)$ is specified as (4.11) and have been proposed by Battese and Coelli (1992):

$$G(t) = \exp\left[\gamma(t - T)\right], \qquad\qquad (4.11)$$

where T is the terminal period of the sample. As Model 5, in this model the inefficiency is time-driven and the simpler one-parameter function is estimated using ML method. In this model u_i specifies the terminal distribution of inefficiency because $u_{it} = u_i$ when t=T (as $G(t) = \exp[\gamma(t - T)] = 1$)). Additionally, when $\gamma \prec 0$, the model predicts convergence in inefficiency to u_i.

4.2.2.4   Model 7

Kumbhakar and Wang (2005) used the following specification of G(t) to specify a model of time-variant efficiency driven by time:

$$G(t) = \exp\left[\gamma\left(t - \underline{t}\right)\right], \tag{4.12}$$

where $\underline{t}$ is the beginning period of the sample. This is the opposite of Model 6 where T represents the last period of observation. The reference points in these two models are the initial and final periods. Analytically Models 6 and 7 are the same, but they are interpreted differently. Kumbhakar (1990) and its reformulation in Battese and Coelli (1992), $u_i \sim N^+\left(\mu, \sigma_u^2\right)$ specifies the distribution of inefficiency at the terminal point, that is, $u_{it} = u_i$ when $t = T$. With (4.11), $u_i \sim N^+\left(\mu, \sigma_u^2\right)$ specifies the initial distribution of inefficiency. The strength of Model 6 is in accounting for market entry, while $\underline{t}$ accounts for market exit in formulating the reference point in Model 7. A mixture model formulation of the two initial and terminal reference points might be superior to the two models individually. An accounting for entry and exit influence the time variant reference efficiency points as well as improves the correspondence between sample and population in situations where the population structure is changing fast over time. The coefficient γ in the G(t) function has a straightforward interpretation. Since $\gamma = \partial \ln u_{it}/\partial t$, one can interpret it as the percentage change in inefficiency over time. Thus, if γ is negative, technological catch-up and movement towards the frontier is observed and $-\gamma$ is defined as the technological catch-up rate. The γ coefficient is also related to the rate of convergence or change in inefficiency (see Kumbhakar and Wang 2005).

4.2.3   Models Separating Inefficiency and Unobserved Individual Effects

4.2.3.1   Models 8 and 9

The model as specified in (4.1) and (4.2) is a standard panel data model where α_i is an unobservable individual effect. Standard panel data fixed and random-effects estimators are applied to estimate the model's parameters (including α_i). The only difference is that we transform the estimated value of $\hat{\alpha}_i$ to obtain the estimated value of u_i, namely $\hat{u}_i$, by using the highest $\hat{\alpha}_i$ as a reference for the frontier.

A notable drawback of this method is individual heterogeneity cannot be distinguished from inefficiency effects. In other words, all time-invariant heterogeneity, for example soil quality in agricultural production, that is not necessarily inefficiency components is included as inefficiency and therefore $\hat{u}_i$ might be picking up heterogeneity in addition to or even instead of inefficiency (Kumbhakar and Heshmati

1995; Greene 2005b). Outliers serving as a reference and confounded inefficiency can overestimate or bias performance estimates.

Another potential issue with Models 1 and 2 is the time-invariant assumption of inefficiency. If T is large, it seems implausible that the inefficiency of a firm will stay constant for an extended period of time and that a firm with persistent inefficiency will survive in the market. So should one view the time-invariant component as persistent inefficiency or as individual heterogeneity that captures the effects of time-invariant covariates and has nothing to do with inefficiency? If the latter is true, then the results of the time-invariant inefficiency models are wrong. Perhaps the truth lies somewhere in between, that is, a part of the inefficiency might be persistent, while another part may be transitory. Unless the parts are separated from time-invariant individual effects, one has to choose either the model in which α_i represents persistent inefficiency or the model in which α_i represents an individual-specific effect (heterogeneity). Following Kumbhakar and Heshmati (1995) we consider both specifications in this chapter. Thus, the models can be written as:

$$y_{it} = \alpha_i + x'_{it}\beta + v_{it} - u_{it} \tag{4.13}$$

If α_i as treated as fixed parameters which are not part of inefficiency, then the model becomes the so called 'true fixed-effects' panel stochastic frontier model (Model 8) (Greene 2005a). On the other hand, model (4.13) is labeled as 'true random-effects' panel stochastic frontier model when α_i is treated as a random variable (Model 9).[2]

4.2.3.2 Model 10

Using a different approach, Wang and Ho (2010) solved the problem in Greene (2005a) by proposing a stochastic frontier model in which the within and first-difference transformation of the model can be carried out and yet a closed-form likelihood function can still be obtained using the standard practice used in literature. The Wang and Ho (2010) model is written as:

$$y_{it} = \alpha_i + x'_{it}\beta + \epsilon_{it}$$
$$\epsilon_{it} = v_{it} - u_{it},$$
$$v_{it} \sim N(0, \sigma_v^2),$$
$$u_{it} = h_{it}u_i$$
$$h_{it} = f(z'_{it}\delta),$$
$$u_i \sim N^+(\mu, \sigma_u^2), \tag{4.14}$$

---

[2]For more details please see Greene (2005a, b).

In this setup, h_{it} is a positive function of a $1 \times L$ vector of non-stochastic ineffi-ciency determinants (z_{it}) and z_{it} is a vector of variables explaining the inefficiency. The key feature of Model 10 is the multiplicative form of inefficiency effects (u_{it}) in which individual-specific effects, u_i, appear in multiplicative forms with individ-ual and time-specific effects, h_{it}. As u_i^* does not change with time, the within and first-difference transformations leave this stochastic term intact.

4.2.4 Models Separating Persistent Inefficiency from Unobservable Individual Effects

4.2.4.1 Model 11

Time-variant models (Model 4 through Model 11) fail to capture persistent ineffi-ciency, which is hidden within firm effects. Therefore, these models are mis-specified and tend to produce a downward bias in an estimate of overall inefficiency, especially if persistent inefficiency exists and its magnitude is significant. In this regard, Models 11 and 12 which separate the persistent and time-varying inefficiency components are introduced.

Identifying the magnitude of persistent inefficiency is important, especially in short panels, because it reflects the effects of inputs like management (Mundlak 1961) and other unobserved inputs which vary across firms but not over time. The residual component of inefficiency might change over time without any change in a firm's operations. Therefore, a distinction between the persistent and residual components of inefficiency is important and they also have different policy implications. Thus, Model 11 is specified in line with Kumbhakar and Heshmati (1995) as:

$$y_{it} = \beta_0 + x_{it}'\beta + \in_{it}$$
$$\in_{it} = \upsilon_{it} - u_{it}$$
$$u_{it} = u_i + \tau_{it} \tag{4.15}$$

where τ_{it} is the residual (time-varying) component of technical inefficiency which is both firm- and time-specific. u_i is the persistent component (for example, time-invariant management effects) and is only firm-specific. τ_{it} and u_i both are non-negative. To estimate the model (4.15) is rewritten as:

$$y_{it} = \alpha_i + x_{it}'\beta + \omega_{it}$$
$$A_i \equiv \beta_0 - u_i - E(\tau_{it}) \, and$$
$$\omega_{it} = \upsilon_{it} - (\tau_{it} - E(\tau_{it})) \tag{4.16}$$

where ω_{it} is error component and have zero mean and constant variance. The model can be estimated either by generalized least squares (GLS) method or by least squares dummy variable (LSDV) approach. Following Kumbhakar and Heshmati (1995) we use a multi-step procedure to estimate the model. In step 1, we estimate (4.16) using the standard fixed-effects panel data model to obtain consistent estimates of β. In step 2, persistent technical inefficiency is estimated, u_i. In step 3, we estimate β_0 and the parameters associated with the random components, v_{it} and τ_{it}. Finally, in step 4, we estimate the time-varying (residual) component of inefficiency (τ_{it}). The multi-step procedure is cumbersome, but has the advantage of avoiding strong distributional assumption by estimating the model using the ML estimation method.

4.2.4.2 Model 12

Because the assumptions made in previous models are not fully satisfactory, we introduce a final model by Kumbhakar et al. (2014), Colombi et al. (2014) that overcomes some of the limitations of the earlier models. Model 12 is a modified and extended version of Model 11 and includes random firm effects. In this model the error term is split into four components which given the inputs take into account different factors affecting output. The first component captures firms' latent heterogeneity (Greene 2005a, b), which has to be disentangled from inefficiency effects; the second component captures short-run (time-varying) inefficiency. The third component captures persistent or time-invariant inefficiency as in Kumbhakar and Hjalmarsson (1993, 1995), Kumbhakar and Heshmati (1995), while the last component captures random shocks.

Therefore, the final parametric model of this chapter is specified as:

$$y_{it} = \alpha_0 + f(x_{it}; \beta) + \mu_i + v_{it} - \eta_i - u_{it} \tag{4.17}$$

where μ_i are random firm effects that capture unobserved time-invariant inputs and v_{it} is noise. η_i and u_{it} are inefficiency component of the model.

Estimation of the model in (4.17) can be undertaken in a single stage MLE method based on the distributional assumption of the four components (Colombi et al. 2011). However, here a simpler multi-step procedure is considered. For this, the model in (4.17) is rewritten as:

$$y_{it} = \alpha_0^* + f(x_{it}; \beta) + \alpha_i + \varepsilon_{it} \tag{4.18}$$

where $\alpha_0^* = \alpha_0 - E(\eta_i) - E(u_{it})$; $\alpha_i = \mu - \eta_i + E(\eta_i)$; and $\varepsilon_{it} = v_{it} - u_{it} + E(u_{it})$. This model can be estimated in three steps. In the first step, the standard random-effect panel regression is used to estimate $\hat{\beta}$. This procedure also gives predicted values of α_i and ε_{it}, which we denote by $\hat{\alpha}_i$ and $\hat{\varepsilon}_{it}$. Time-varying technical inefficiency, u_{it}, is estimated in the second step and in the final step, we can estimate η_i following a procedure similar to that in step 2 (Jondrow et al. 1982).

Presistent technical efficiency can then be estimated from $PTE = -exp(\eta_i)$. The overal technical efficiency, OTE, is then obtained from the product of PTE and RTE, that is, $OTE = PTE \times RTE$ (Kumbhakar et al. 2015).

In order to provide an overview of the parametric models' structures and differences, Table 4.1 gives a summary of the models based on their common characteristics related to the error component's structure, treatment of firm-specific and time-specific effects, technical efficiency and its temporal structure and the estimation methods and their underlying assumptions.

4.3 Non-parametric Models

This part of chapter four introduces 4 non-parametric models (two full frontier and two partial frontier models) which are used to study the performance of different cotton and sugar beet producing provinces in Iran.

4.3.1 Full Frontiers

4.3.1.1 Model 13 (DEA)

Data Envelopment Analysis (DEA) method is a non-parametric mathematical programming approach used for evaluating a set of comparable decision-making units; it measures productive efficiency of the DMUs (Danijela 2012). It is a full frontier method which estimates the production frontier and evaluates the technical efficiency of each DMU.

Charnes et al. (1978) have presented the following model to measure the efficiency of DMU k non-parametrically:

$$\min \theta$$
$$\text{Subject to } \theta X_{ik} - \sum_{j=1}^{n} \lambda_j X_{ij} \geq 0, \; i = 1, \ldots, m.$$
$$\sum_{j=1}^{n} \lambda_j Y_{rj} \geq Y_{rk}, \qquad r = 1, \ldots, s. \qquad (4.19)$$
$$\sum_{j=1}^{N} \lambda_j = 1, \lambda_j \geq 0, \qquad j = 1, \ldots, n.$$

For DMU k, Y_k and X_{ik} denote the level of the output and the level of the ith input, respectively. The optimal level of θ, denoted by θ^*, satisfies the condition $0 \leq \theta^* \leq 1$. If θ equals one, the DMU under measurement lies on the estimated frontier and is said to be technically fully efficient.

In (4.19) the constant returns to scale of the production function are assumed, therefore this model is often referred to as the CCR (Charnes et al. 1978) model. The obtained scores of the CCR model determine technical efficiency and distinguish it from other types of efficiencies (such as allocative efficiency) in which no costs and prices are used (Yang and Chang 2009).

Table 4.1 Main characteristics of different models

	Model 1	Model 2	Model 3	Model 4	Model 5	Model 6	Model 7	Model 8	Model 9	Model 10	Model 11	Model 12
General firm effects	No	No	No	Fixed	Fixed	Fixed	No	Fixed	Random	Fixed	No	Random
Technical inefficiency components:												
Persistent	No	No	No	No	No	No	No	No	No	No	Yes	Yes
Residual	No	No	No	No	No	No	No	No	No	No	Yes	Yes
Overall technical inefficiency:												
Mean	Time-inv.	Time-inv.	Time-inv.	Time-inv.	Time-inv.	Time-inv.	Time-inv.	Zero trunc.	Zero trunc.	Zero trunc.	Zero trunc.	Zero trunc.
Variance	Homo.	Homo.	Homo.	Hetero.	Hetero.	Hetero.	Homo.	Hetero.	Homo.	Homo.	Homo.	Homo.
Symmetric error term:												
Variance	Homo.	Homo.	Homo.	Hetero.	Hetero.	Hetero.	Homo.	Homo.	Homo.	Homo.	Homo.	Homo.
Estimation Method:	COLS	GLS	ME	OLS	MLE	MLE	MLE	MLE	MLE	MLE	MLE	MLE

Notes Fixed effects (Fixed), Random effects (Random), Homoscedastic variance (Homo.), Time invariant efficiency (Time inv.), Zero truncated error term (Zero trunc.), Corrected ordinary least squares (COLS), Maximum likelihood Estimation (ML), and Generalized least squares (GLS)

4.3.1.2   Model 14 (FDH)

Another method which has received a considerable amount of research attention in efficiency measurement literature is the Free Disposable Hull (FDH). This model was introduced by Deprins et al. (1984). The FDH estimator is both a deterministic and a non-parametric method for technical efficiency measurement. It is deterministic since it cannot accommodate stochastic properties. On the other hand, The FDH estimator is non-parametric because it lacks functional form specifications. The most important advantage of FDH is that efficiency evaluations are affected only by the actually observed performance (Subhash 2004). But as DEA, FDH had a disadvantage that it is very sensitive to noise and outliers. It is also susceptible to dimensionality problems (Gabdo et al. 2014).

The FDH estimator only imposes free disposability of inputs; it does not impose convexity of the estimated technology (Silva et al. 2016). Following DeBorger et al. (1994), FDH is derived as:

Suppose $Y = Y(Y_1, Y_2, \ldots, Y_n)$ presents n non-negative outputs produced by m inputs $X = X(X_1, X_2, \ldots, X_m)$. Then, the FDH estimator is defined by the following axioms:

Axiom I: $0 \; L(Y)$ for $Y \geq 0$, and $L(0) = R_+^n$

Axiom I assumes that it is not possible to obtain a semi-positive output from a null input vector. Axiom II states that for any utilization of finite inputs, finite outputs are produced:

Axiom II: *if* $\left| Y^l \right| \rightarrow +\infty$ as $l \rightarrow +\infty$, then $\cap_{l=1}^{+\infty} L(Y^l)$ is empty

Then we have the axiom of free disposability of inputs (Axiom III), which implies that an increase in input X cannot lead to a decrease in output Y:

Axiom III: if $X \in L(Y)$ and $X' \geq X$, then $X' \in L(Y)$

Axiom IV: $L(y)$ is a closed correspondence

Axiom IV indicates that if an array of input vectors can each yield output bundle Y and converge to X^*, then the same X^* can also yield output bundle Y.

Axiom V: if $Y' \geq Y$, then $L(Y') \subseteq L(Y)$

Strong free disposability of output (Axiom V) provides for variable returns to scale and assumes any reduction in output with the same quantity of inputs. Therefore, the specification of the FDH input correspondence is:

$$L(Y)^{FDH} = \{X \mid X \in R_+^m, Z'J \geq Y, \; Z'N \leq X, \; I_k'Z = \Big|, \; Z_i \in \{0, 1\}\} \qquad (4.20)$$

where J represents the $k \times n$ matrix of outputs, N represents the $k \times 1$ vector of intensity, I_k indicates the $k \times 1$ vector of ones. According to all axioms, convexity assumption is not imposed on technology.

4.3.2 Partial Frontiers

Despite the advateges of using full frontier models, one important disadvantage may arise from using DEA and FDH, namely, is the so-called "Curse of dimensionality". It is related to problems associated with a low number of Decision Making Units relative to the number of input-output variables. In this regard, some studies have evolved to provide soultions to this problem. Order-m and Order-α estimtors are roubust undicators to the "Curse of dimensionality" and also to the presence of outliers in the data. Other advantage of Order-m and Order-α estimtors is that non of them require convexity assumption.

4.3.2.1 Model 15 (Order-m)

Order-m is a generalization of FDH model and is a result of adding a layer of randomness to the computation of efficiency scores in the FDH model (Daraio and Simar 2007; Cazals et al. 2002). It introduces a benchmark frontier which is less sensitive to extreme observations as compared to the full frontier. This benchmark is defined as the expected minimal input value among m ($m \geq 1$) peers:

$$Q_m(y) = E[\min(X_1, \ldots, X_m)|Y \geq y], \qquad (4.21)$$

where $Q_m(y)$ is the minimal input frontier function. Then we have the equivalences, in which as $m \to \infty$, $Q_m(y) \to Q(y)$:

$$Q_m(y) = \int_0^\infty S^m(u|y)du = Q(y) + \int_{Q(y)}^\infty S^m(u|y)du. \qquad (4.22)$$

By plugging the empirical version of $S(u|y)$ in (4), a non-parametric estimator of $Q_m(y)$ is given as:

$$\hat{Q}_m(y) = \int_0^\infty \hat{S}^m(u|y)du. \qquad (4.23)$$

According to Cazals et al. (2002), for fixed m we have: $\sqrt{n}(\hat{Q}_m(.) - Q_m(.)) \xrightarrow{\lambda} \xi(0, \Omega)$, ($\xi$ is a Gaussian process with covariance function Ω). Therefore, for a fixed value of m and any given y, as $n \to \infty$, we have:

$$\frac{\sqrt{n}}{\sigma^2(m, y)}(\hat{Q}_m(y) - Q_m(y)) \xrightarrow{\lambda} N(0, 1) \qquad (4.24)$$

where:

$$\sigma^2(m, y) = E\left[\frac{m\psi(Y \geq y)}{S_Y(y)} \int_0^\infty (S^{m-1}(u|y)\psi(X \geq u) - S_m(u|y))du\right]^2 \qquad (4.25)$$

Therefore, $\hat{Q}_m(y)$ will converge to the FDH estimator ($\hat{Q}(y)$), as m $\rightarrow \infty$.[3]

4.3.2.2 Model 16 (Order-α)

The other concept of a partial frontier model mentioned earlier is the Order-α quintile-type frontier, which provides a robust estimator of the frontier function (Daouia and Simar 2005; Aragon et al. 2005). The idea behind Order-α quintile-type frontier is to determine the frontier by first fixing the probability (1-α) of observing points below this Order-α frontier. Order-α reverse the causation of Order-m and choose the proportion of the data lying directly below the frontier.

According to Tauchmann (2011) Order-α generalizes the FDH estimator by employing the (100-α)[th] percentile approach. Order-α minimizes input consumption among available peers for benchmarking. This model is written as:

$$\hat{\theta}_{\alpha i} = {}^{P(100-\alpha)}_{j \in B_i} \left\{ \max_{k=1,...,K} \left\{ \frac{x_{kj}}{x_{ki}} \right\} \right\} \tag{4.26}$$

If $\alpha = 100$, both Order-α and FDH result in the same output, while for values of $\alpha \prec 100$ some super-efficient DMUs may result. Also for $\alpha \prec 100$ some observations might be un-enveloped by the estimated production possibility frontier, which are called super-efficient DMUs (Gabdo et al. 2014).

In Order-m and Order-α models we need to choose values for parameters m and α. Different amounts of m and α define the position of the frontier relative to the data. Therefore, the choice of m and α is critical. m represents the size of the artificial reference sample and the default is (m = ceil ($N^{2/3}$)), where ceil (.) stands for the ceiling function. Only integer and positive amounts are allowed for m values (Tauchmann 2011). As the value of m and α increases the number of data used in the estimation increases. Therefore, efficiency scores are dependent on the choice of m and α. The choice of practical values for m and α is a theoretical issue.

4.4 Window Analysis

In the next chapter of this book we plan to empirically measure the technical efficiency of different provinces in Iran not only for a single year but for a time period of 2000–2012, which is considered a dynamic evaluation and could provide us with more information about the efficiency changes. Therefore, it is practical to explore it by applying window analysis.

Window analysis is a variation of the traditional non-parametric approach, and handles cross-sectional and time-varying data in order to measure dynamic effects. This technique is based on the principle of moving average and establishes efficiency

---

[3]For more details please see Cazals et al. (2002).

measures by treating each DMU in different periods as a separate unit. Under the window analysis framework, performance of a province in a period can be contrasted to the performance of other provinces as well as to its own performance in other periods. Therefore, by applying this technique we can evaluate technical efficiency of different provinces in different years through a sequence of overlapping windows. This technique operates on the principle of moving average. Window analysis for technical efficiency measured in this study is presented below.

A window with $m \times s$ observations is denoted starting at time t $(1 \le t \le T)$ with window width s $(1 \le s \le T - t)$. The window width is supported by the number of time periods (years in this study) under analysis, and the time periods are conceived of in an inter-period manner. Since each of the provinces for a specific year within a given window is measured against each other, window analysis implicitly assumes that there are no technical changes during the period under analysis within each window. One can use Windows with narrow with, in order to relieve this problem (Zhang et al. 2011). According to Charnes et al. (1994), (Wang et al. 2013) a window width of three or four time periods tends to yield the best balance of stability of the efficiency measure.

As an example, in cotton production, 13 provinces and a time period of 13 years of efficiencies needs to be examined, so $m = 13$ and $T = 13$. Also based on the data structure we decided to choose the window width to be four years as optimal. Therefore, the first four years of 2000, 2001, 2002 and 2003 construct the first window. Then the window moves on a one-year period by dropping the original year and adding a new year. This process continues until the last window, which contains the last four years of 2009, 2010, 2011 and 2012, is constructed. At last, we obtain ten window which are performed for each DMU and the number of DMUs (provinces in our study) in each window becomes 52 ($m \times s = 13 \times 4$).

Chapter 5
Models Applied to Iran's Cotton and Sugar Beet Production

Abstract In this chapter all the parametric and non-parametric models are applied to two datasets from Iran's cotton and sugar beet producing provinces. The data for cotton crop is a balanced panel of 13 provinces observed over 13 years from 2000 to 2012, with 169 observations. The dataset for sugar beet is also a balanced panel and includes 143 observations from 11 provinces over 13 years. The main variables used in the different models include output, labor, seeds, pesticides, chemical fertilizers and animal fertilizers. The variables which influence technical efficiency are share of chemical fertilizers in total fertilizers and the machinery utilization rate measured in percentage use. This chapter also sheds light on whether particular panel data stochastic frontier models are better suited to different datasets. It conducts tests of functional forms and nestedness and analyzes them. Based on these criteria and tests the best parametric model for each dataset is selected. The most efficient and inefficient provinces in cotton and sugar beet production are recognized based on the most suitable model and technical efficiency scores. Finally, the most efficient provinces in cotton and sugar beet production of Iran are recognized.

5.1 Introduction

This chapter includes the estimation results of the parametric and non-parametric models introduced in the study. It presents the results of cotton crop in the first section by first explaining the descriptive results and statistical characteristics of the output and input variables. It then presents the estimation results for 12 frontier production functions and four non-parametric models along with technical efficiency measurement. The second section gives details of the sugar beet crop.

Some parametric models of this chapter are discussed in Rashidghalam et al. (2016).

© Springer Nature Singapore Pte Ltd. 2018
M. Rashidghalam, *Measurement and Analysis of Performance of Industrial Crop Production: The Case of Iran's Cotton and Sugar Beet Production*, Perspectives on Development in the Middle East and North Africa (MENA) Region, https://doi.org/10.1007/978-981-13-0092-9_5

5.2   Results of Cotton Production

5.2.1   Descriptive Results

This book is based on information about the cotton crop's inputs and outputs in 13 provinces of Iran during 2000–2012. The study provinces are Markazi, Mazandaran, East Azerbaijan, Fars, Kerman, Khorasan, Isfahan, Semnan, Yazd, Tehran, Golestan, Ardabil and Qom. Information on the harvested area, production and yield of cotton during the study period are given in Figs. 5.1, 5.2 and 5.3. Based on Figs. 5.1 and 5.2, Khorasan and Golestan provinces had the largest harvested area in the country. These provinces also had the highest cotton production in the country. Yazd and Mazandaran had the minimum harvested area and the minimum production of this crop among the studied provinces. According to these two figures, the harvested area and the amount of cotton produced in Tehran, Golestan, Ardabil and Qom provinces showed a downward trend during the study period. According to Fig. 5.3, Tehran and Golestan had the highest and the lowest yield of cotton production respectively among the provinces. As seen in the figure, the yields of cotton producing provinces had severe and irregular fluctuations, but what is almost the same for all provinces was a sharp drop in 2010. Probably one of the reasons for this drop in cotton yield in Iran was the drought. In 2010, Iran experienced one of its least rainy and driest years. It began with a dry autumn with the result that at the end of autumn, severe drought had covered more than 97% of the country's land area. In this season, precipitation decreased three times (18.5–60 mm long-term rainfall).

Table 5.1 gives the statistical characteristics of the variables used in this study. It shows that the average yield of cotton production in Iran was about 2,445 kg per hectare. The highest cotton yield was in Tehran province, followed by East Azerbaijan and Isfahan. The minimum yield was in Golestan province (with an average of 1,914 kg/ha) which varied from a minimum of 1,410 kg to a maximum of 2,392 kg per hectare.

The highest and lowest amount of pesticides used were in Mazandaran (9.18 kg/ha) and Tehran (1.08 kg/ha) respectively. Pesticides used in the production of cotton include herbicides, insecticides and fungicides. The average consumption of pesticides during the study period in the 13 provinces was 4.59 kg/ha. According to Table 5.1, about 80 labor per hectare was used to produce cotton in Iran and the highest labor use was in Isfahan and Yazd provinces. Fertilizers are another input used to estimate the frontier production function. The average use of this input was 439 kg/ha with Qom being its largest consumer. Fertilizers used in cotton production include phosphate, nitrogen and potassium. The fourth input used in this study is animal fertilizers with an average consumption of 1,740 kg/ha. Fars province had the lowest consumption of this input.

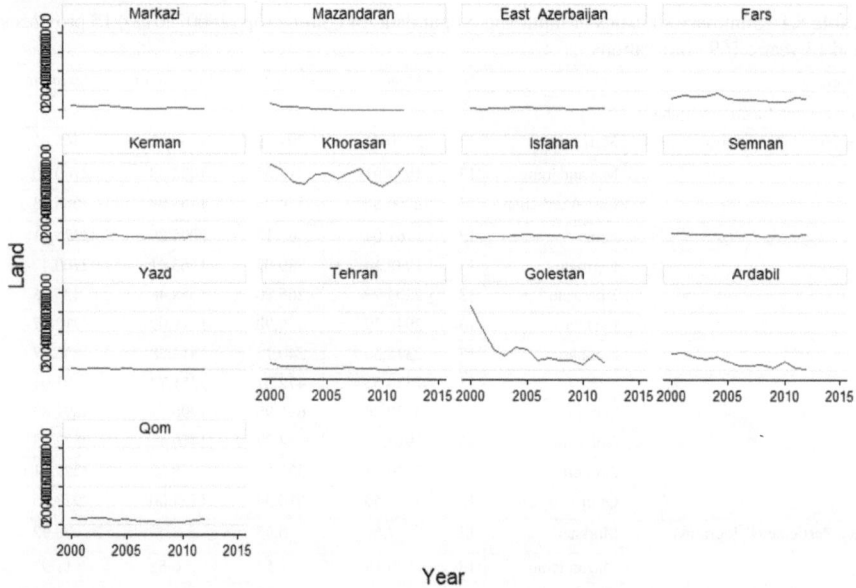

Fig. 5.1   Cotton harvested area in different provinces (2000–2012)

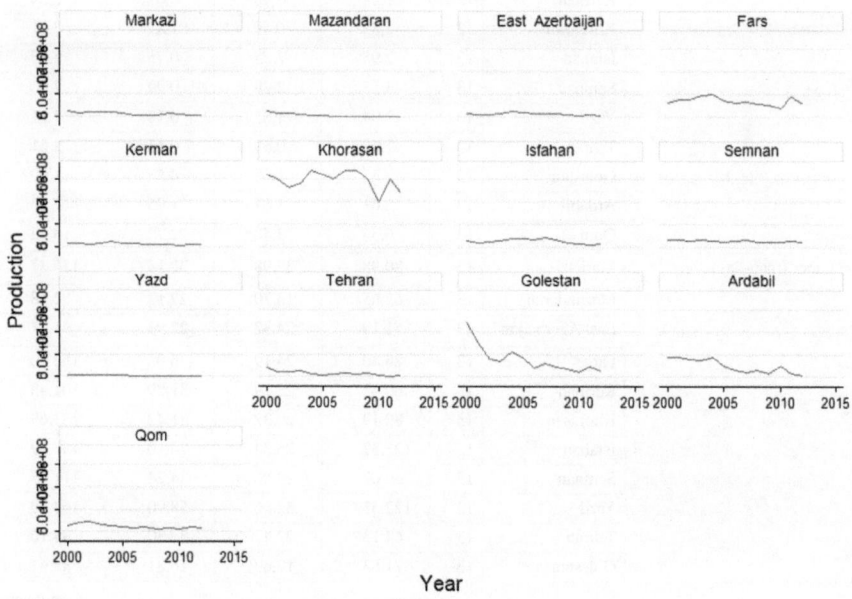

Fig. 5.2   Cotton production in different provinces (2000–2012)

Table 5.1 Summary statistics of input and output data for cotton crop (2000–2012), 13 provinces and 13 years, 169 observations

Variable	Province	N	Mean	Std. dev.	Minimum	Maximum
Production function variables:						
y Output (Kilograms)	Markazi	13	2370.61	294.07	1662.93	2765.05
	Mazandaran	13	1934.91	414.35	1092.38	2700.00
	East Azerbaijan	13	2956.56	554.26	1788.99	3666.73
	Fars	13	2761.04	368.12	2095.89	3375.01
	Kerman	13	1938.14	389.03	1165.41	2700.47
	Khorasan	13	2204.79	265.20	1558.46	2425.96
	Isfahan	13	2948.70	536.98	1620.09	3712.77
	Semnan	13	2375.36	240.13	1813.28	2856.27
	Yazd	13	2384.53	472.01	1351.72	2891.01
	Tehran	13	3120.28	695.95	1796.74	4894.89
	Golestan	13	1914.77	284.77	1410.41	2392.23
	Ardabil	13	2646.73	351.56	1928.83	3229.41
	Qom	13	2232.59	510.34	1234.00	2978.96
x_p Pesticide (Kilograms)	Markazi	13	7.69	6.75	3.40	28.92
	Mazandaran	13	9.18	1.54	6.52	11.93
	East Azerbaijan	13	7.03	3.77	1.41	13.87
	Fars	13	4.90	1.46	2.17	7.56
	Kerman	13	1.49	0.40	0.61	2.04
	Khorasan	13	2.53	*0.59*	1.81	3.61
	Isfahan	13	2.95	*1.83*	0.75	7.95
	Semnan	13	3.57	2.28	0.99	7.80
	Yazd	13	2.04	1.85	0.18	7.11
	Tehran	13	1.08	0.69	0.15	2.61
	Golestan	13	7.33	1.58	4.54	11.09
	Ardabil	13	3.61	1.67	1.25	7.93
	Qom	13	6.34	3.48	2.72	13.62
x_l Labor (man-day)	Markazi	13	80.98	32.98	38.13	156.37
	Mazandaran	13	65.05	13.70	27.13	82.18
	East Azerbaijan	13	51.19	24.32	22.93	91.77
	Fars	13	86.40	24.19	46.08	152.20
	Kerman	13	60.43	22.29	31.29	101.43
	Khorasan	13	82.19	20.32	41.42	133.69
	Isfahan	13	123.37	25.34	75.10	157.29
	Semnan	13	60.07	29.06	24.24	111.08
	Yazd	13	122.53	32.66	58.90	162.60
	Tehran	13	64.13	22.47	32.30	94.10
	Golestan	13	71.63	17.69	16.81	85.62

(continued)

Table 5.1 (continued)

Variable	Province	N	Mean	Std. dev.	Minimum	Maximum
	Ardabil	13	62.67	17.44	28.00	90.99
	Qom	13	70.18	19.12	34.16	108.74
x_{cf} Chemical fertilizers (Kilograms)	Markazi	13	465.50	91.85	337.57	692.67
	Mazandaran	13	256.60	14.20	233.33	286.95
	East Azerbaijan	13	341.30	116.02	161.09	602.27
	Fars	13	512.88	84.39	323.32	654.89
	Kerman	13	345.18	77.37	180.42	470.75
	Khorasan	13	412.38	68.45	253.28	551.37
	Isfahan	13	583.21	173.35	262.48	925.03
	Semnan	13	424.39	130.37	218.38	601.43
	Yazd	13	558.04	148.27	268.61	890.00
	Tehran	13	412.14	72.58	276.91	569.63
	Golestan	13	373.00	290.41	247.90	1332.32
	Ardabil	13	344.84	58.81	186.20	437.40
	Qom	13	679.92	184.09	455.37	1107.89
x_{af} Animal fertilizers (Kilograms)	Markazi	13	222.34	99.89	9.00	439.00
	Mazandaran	13	456.00	165.75	50.00	862.00
	East Azerbaijan	13	241.00	0.00	241.00	241.00
	Fars	13	167.50	379.78	11.00	1400.00
	Kerman	13	987.80	523.20	479.15	2204.69
	Khorasan	13	1191.22	518.49	514.90	2185.00
	Isfahan	13	4142.39	6273.52	398.00	24228.00
	Semnan	13	2047.01	2975.91	128.00	11626.00
	Yazd	13	2080.57	1391.37	323.00	4001.00
	Tehran	13	5968.56	6076.70	30.00	22757.00
	Golestan	13	196.09	239.73	2.00	937.00
	Ardabil	13	613.28	871.50	31.00	3424.33
	Qom	13	4314.31	3887.86	575.00	14632.90
Inefficiency determinant variables in inefficiency function:						
Z_1 Rate of chemical fertilizers per total fertilizers	Markazi	13	0.69	0.12	0.49	0.98
	Mazandaran	13	0.39	0.14	0.23	0.85
	East Azerbaijan	13	0.57	0.08	0.40	0.71
	Fars	13	0.85	0.22	0.19	0.98
	Kerman	13	0.29	0.11	0.12	0.46
	Khorasan	13	0.28	0.08	0.16	0.42
	Isfahan	13	0.23	0.16	0.04	0.66
	Semnan	13	0.32	0.23	0.02	0.80
	Yazd	13	0.30	0.20	0.07	0.65
	Tehran	13	0.21	0.28	0.02	0.94
	Golestan	13	0.70	0.22	0.24	0.99
	Ardabil	13	0.48	0.19	0.08	0.92
	Qom	13	0.23	0.17	0.04	0.57

(continued)

Table 5.1 (continued)

Variable	Province	N	Mean	Std. dev.	Minimum	Maximum
Z_2 machinery (per cent)	Markazi	13	39.55	10.07	27.75	63.33
	Mazandaran	13	37.23	5.59	25.00	50.10
	East Azerbaijan	13	40.51	17.47	20.20	76.01
	Fars	13	37.59	4.36	31.99	46.25
	Kerman	13	43.86	6.55	30.64	56.25
	Khorasan	13	35.57	3.27	30.19	41.39
	Isfahan	13	21.35	5.91	11.01	35.26
	Semnan	13	40.99	11.34	28.77	67.40
	Yazd	13	23.09	8.31	7.16	35.20
	Tehran	13	45.50	4.52	39.95	57.42
	Golestan	13	38.55	7.72	27.55	57.00
	Ardabil	13	58.36	9.79	36.67	72.96
	Qom	13	39.96	4.86	33.61	51.52

Source: Ministry of Agriculture Jihad, 2000–2012

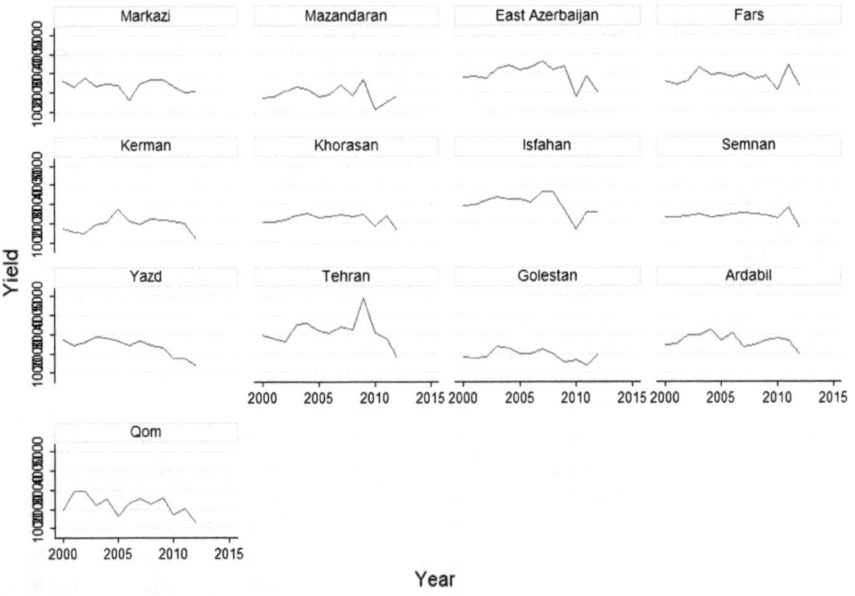

Fig. 5.3 Cotton yield in different provinces (2000–2012)

The second part of Table 5.1 gives information on the determinants of inefficiency. Studies on technical inefficiency discuss the impact of various factors such as management, manager's age, experience, education, participation in extension classes and the use of machinery on efficiency. However, since we did not have access to these variables at the provincial level only two variables—the ratio of chemical fertilizers to total fertilizers (fertilizers + animal fertilizers) (Z_1) and the percentage of machinery used per hectare (Z_2)—were used. According to Table 5.1, the average ratio of fertilizers to total fertilizers was 42%. On the other hand, the average percentage of machine use per hectare was 38%; Ardebil and Isfahan provinces had the highest and lowest percentage respectively.

5.2.2 Parametric Models

5.2.2.1 Results of Stochastic Frontier Models

We analyzed a flexible translog production functional form for Models 1–12. A Cobb-Douglas (CD) production function was also estimated as a special and restricted case of the translog function. According to the CD production function's results this functional form imposed restrictions on technology by restricting production elasticities as constant and the elasticity of input substitution to be *unity*. We tested this function against the translog function and conclude that the general translog model was more appropriate than the restricted CD form for presenting our data.

When using the specification of a translog production function the first-order coefficients are not conclusive because they do not provide much information on the responsiveness of the output to changes in the use of various inputs. If the variables in the translog model are normalized at the sample mean, then the first-order coefficients are estimates of elasticities at mean input levels. Consequently, we used the first-order coefficients of input variables in the translog model to calculate output elasticity with respect to each input in the production function at their mean values. By using the mean-scaled variables it was possible to interpret the first-order coefficients of the translog function as the partial elasticities of production for the sample mean. They reflect the total effects of changes in an input on changes in output, everything else being given. It should be noted that mean normalized based elasticities are crude measures and are not really representative of the provincial variations in the elasticities and are not weighted. The estimation results for cotton production are given in Table 5.2.

Table 5.2 Estimated stochastic frontier models, N = 169 observation

	Model 1	Model 2	Model 3	Model 4	Model 5	Model 6	Model 7	Model 8	Model 9	Model 10	Model 11	Model 12
x_p	0.082[c]	0.071[b]	0.080[b]	0.058	0.086[a]	0.095[a]	0.096[a]	0.113[a]	0.076[a]	0.090[b]	0.082[c]	0.071[a]
x_{cf}	0.517[a]	0.633[a]	0.542[a]	0.476[a]	0.514[a]	0.530[a]	0.531[a]	0.394[a]	0.548[a]	0.436[a]	0.517[a]	0.633[a]
x_l	0.260[a]	0.221[a]	0.261[a]	0.177[a]	0.223[a]	0.191[a]	0.192[a]	0.259[a]	0.257	0.254[a]	0.264[a]	0.221[a]
x_{af}	0.009	0.013	0.012	0.014	0.028[c]	0.036[c]	0.036[c]	0.087[a]	0.016[a]	0.086[b]	0.009	0.013
$(1/2)x_p^2$	0.067	0.037	0.059	0.062	0.047	0.038	0.038	0.050	0.049[a]	0.068	0.067	0.037
$(1/2)x_{cf}^2$	0.067	-0.186	0.007	0.128	-0.035	-0.059	-0.059	-0.037	0.002	0.055	0.045	-0.186
$(1/2)x_l^2$	0.045	0.037	0.009	-0.009	0.090	0.048	0.048	0.090	0.02	0.010	0.003	0.037
$(1/2)x_{af}^2$	0.003	-0.006	0.005	0.003	0.008	0.011	0.011	0.034[b]	0.008	0.034	0.005	-0.006
$(x_p)*(x_{cf})$	0.005[c]	-0.066	-0.100	-0.14[a]	-0.074	-0.070	-0.070[b]	-0.074	-0.089[a]	-0.106[c]	-0.114[c]	-0.066
$(x_p)*(x_{af})$	-0.114	-0.009	0.022	0.055[a]	0.031	0.034	0.034[c]	0.023	0.023[a]	0.022	0.027	-0.009
$(x_{cf})*(x_l)$	0.027	0.149	0.086	0.083	0.044	0.082	0.082	0.035	0.074[a]	0.072[b]	0.080	0.149
$(x_{cf})*(x_{af})$	0.080	-0.018	-0.057	-0.105[a]	-0.054[c]	-0.063	-0.063[b]	-0.059[c]	-0.057[a]	-0.084	-0.065	-0.018

(continued)

Table 5.2 (continued)

		Model 1	Model 2	Model 3	Model 4	Model 5	Model 6	Model 7	Model 8	Model 9	Model 10	Model 11	Model 12
Intercept		16.05[a]	16.04[a]	16.40[a]	15.83[a]	16.33[a]	16.31[a]	16.31[a]	15.95[a]	16.24[a]	–	16.05[a]	16.04[a]
Gamma	Time	–	–	–	–	0.313[a]	–	–	–	–	–	–	–
	Time2	–	–	–	–	-0.038[a]	–	–	–	–	–	–	–
	TimeT	–	–	–	–	–	0.114[a]	0.113[a]	–	–	–	–	–
μ		–	–	0.270	–	-0.378	-1.483	-0.376	–	–	–	–	–
σ_u^2	cons	0.264	0.269	0.084	–	0.235	0.18002	0.076	1.773	0.064	0.264	0.269	–
	Z_1	–	–	–	–	–	–	–	-6.193	–	–	–	–
	Z_2	–	–	–	–	–	–	–	-0.028	–	–	–	–
h	Z_1	–	–	–	–	–	–	–	–	–	-4.796	–	–
	Z_2	–	–	–	–	–	–	–	–	–	0.010	–	–
σ_v^2		0.264	0.267	0.065	–	0.047[a]	-3.037[a]	0.047[a]	0.025	0.042	0.061[a]	0.264	0.267
Returns to scale		0.868	0.938	0.895	0.725	0.851	0.852	0.855	0.853	0.897	0.866	0.872	0.938
R-squared		0.948	0.951	–	0.985	–	–	–	–	–	–	0.948	0.951
Log-likelihood		–0.562	–	–24.56	17.103	–0.630	–58.74	–1.340	9.140	–	–7.310	–0.562	–
AIC		31.124	–	81.129	-8.027	37.261	36.685	37.685	39.717	–	46.612	31.124	–
BIC		71.813	–	131.27	45.337	93.599	89.893	89.893	130.484	–	96.690	71.813	–

Notes: Significant at less than 1% ([a]), 1–5% ([b]), and 5–10% ([c]) levels of significance

The coefficient estimates of the frontier production function were not substantially different across the 12 models. For all the models the estimated output elasticities of inputs such as pesticides, chemical fertilizers and labor were positive and statistically significant at the 5% level indicating that as expected production increased with such inputs. The elasticities for animal fertilizers in some models were also positive but insignificant, indicating that there was scope for increasing cotton production by increasing the use of these inputs in Iran. Elasticity of chemical fertilizers ranged from 0.394 (Model 8) to 0.633 (Model 2) and was significant in all the models. The elasticity of this input in Model groups 1 and 4 was more than 0.5. In the first group all models produced highly similar elasticities. The elasticities for pesticides were between 0.058 and 0.113 for all the models. Group 3 of the models had the highest pesticide elasticity. Elasticity of this input in Model 9 and Model 12 was 0.076 and 0.071 respectively. Estimates of labor elasticity were larger than 0.177 for all models and it was 0.221 in Model 12, meaning that if labor increased by 1%, cotton production will increase by 0.221%. Group 3 of the models had the highest labor elasticity. The sum of these elasticity coefficients was between 0.725 and 0.938, which indicates that the production process had decreasing returns to scale (DRTS). So, if provinces increased all inputs proportionally by 1%, the output will increase by about 0.725 and 0.938%. The scale elasticity was the highest for Model 12 (0.938 on average) and the lowest for Model 4 (0.725 on average). On the other hand, group 4 of the models had the largest scale-economies.

In terms of policy implications, it is more important to determine which variables influence production inefficiency. The determinant variables considered in this study for identifying possible influences on technical efficiency are the share of chemical fertilizers in total fertilizers (z_1) and the rate of machinery utilization (z_2) to indicate the extent to which farms used modern technologies in the production of cotton. The negative sign of the parameters of z variables in Table 5.2 (and Model 8) means that the associated variables had a positive effect on technical efficiency in the production of cotton by reducing inefficiencies. According to Table 5.2 the use of more chemical fertilizers than animal fertilizers led to a significant increase in technical efficiency in cotton production. The results also show that an increased use of machinery led to a significant increase in technical efficiency in Iran's cotton production.

To compare nested and non-nested models we used the Akaike Information Criterion (AIC) and the Bayesian Information Criterion (BIC) and the best suitable input-output relationship was selected using these criteria. AIC and BIC are defined as: $AIC = -2 \ln(\text{maximum likelihood}) + 2k$ and $BIC = -2 \ln(\text{maximum likelihood}) + k \ln(n)$, where k is the number of the estimated parameters and n is the number of observations (Wang and Liu 2006). The models with lowest AIC and BIC values are preferred. Last two rows of Table 5.2 present these criteria for some of the models. Generally speaking, there were a few differences in the selection results of AIC and

Table 5.3 Specification tests for alternative nested models (Model 3 versus Models 4, 5, 6)

	Log- likelihood under H_0	Log- likelihood under H_1	Test statistic	Critical value at 5%	Decision
Model 3 versus Model 4	−24.564	−0.630	47.767	5.135	Model 3 is rejected
Model 3 versus Model 5	−24.564	−58.74	−68.352	2.705	Model 3 is rejected
Model 3 versus Model 6	−24.564	−1.342	46.445	2.705	Model 3 is rejected

BIC. According to Table 5.2, Model 4 (CSS 1990) had the lowest AIC and BIC. Therefore, based on these criteria, Model 4 was preferred to all the other models. Then Model 6 (Battese and Coelli 1992), Model 5 (Kumbhakar 1990) and Model 7 (Kumbhakar and Wang 2005) showed the minimum amount of AIC and BIC. The selection results of BIC were identical to AIC except for Models 8 (Greene 2005a) and 10 (Wang and Hoo 2010), in which based on AIC Model 8 was preferred to Model 10. While according to BIC, Model 10 was preferred to Model 8. Conversely, Model 3 (Pitt and Lee 1981) had the maximum value of AIC and BIC. Therefore, these models were the less favorable models in our dataset. Based on AIC and BIC, Model 4 was the most preferable model in our dataset.

For comparing some nested models we used a generalized likelihood ratio (LR) test. Based on the results of this test (presented in Table 5.3), Model 3 was rejected in favor of the time-variant model (Model 4). Given the results of the statistical tests, Table 5.3 suggests that the time-variant models (Models 5 and 6) were also preferred over the time-invariant models (Model 3).

5.2.2.2   Technical Efficiency Measurement

Descriptive statistics for mean technical efficiency according to different models are presented in Table 5.4. According to this table, unlike elasticities, various models clearly produced different efficiency results. This is due to different assumptions and treatment of the residual and its decomposition across the models. Models 1–3 produced similar technical efficiency. Average technical efficiency was the highest in this group and ranged from 0.71 to 0.74. It compares the production efficiency relative to the best practiced province. The average technical efficiency for Model 4 through Model 7 ranged from 0.46 to 0.48 with large dispersions. The third group of models had the lowest technical efficiency with large variations ranging from 0.24 to

Table 5.4 Descriptive statistics for technical efficiency measures by different parametric models

Technical efficiency	Mean	Std. dev.	Minimum	Maximum
Model 1	0.708	0.177	0.431	1.000
Model 2	0.741	0.153	0.531	1.000
Model 3	0.714	0.152	0.471	0.922
Model 4	0.457	0.198	0.093	1.000
Model 5	0.769	0.182	0.251	0.972
Model 6	0.785	0.183	0.256	0.983
Model 7	0.785	0.183	0.256	0.983
Model 8	0.822	0.152	0.373	0.980
Model 9	0.238	0.114	0.055	0.984
Model 10	0.902	0.116	0.541	0.997
Model 11	0.612	0.157	0.314	0.899
Model 12	0.751	0.097	0.389	0.933

Notes:
Models 1–3: Models with time-invariant inefficiency effects
Models 4–7: Models with time-variant inefficiency effects
Models 8–10: Models separating inefficiency and unobserved individual effects
Models 11–12: Models separating persistent inefficiency from unobserved individual effects

0.90. The mean technical efficiency of Model 10 (0.90) was the highest, while it was the least for Model 9 (0.24). The average technical efficiency of Models 1 and 2 was about 0.71 and 0.74 respectively with a maximum value of 1.0, while the minimum for these models was 0.43 and 0.53 respectively. Therefore, based on Model 1 the gap between the most efficient and inefficient province was about 0.57. In Model 9 the minimum efficiency was 0.055 and the maximum was 0.984. This gap shows that there were large differences in efficiency of cotton production in the different provinces.

Models 1–3 (time-invariant models) produced almost the same technical efficiency scores. The mean value of technical efficiency according to most of the models was more than 0.71. The average efficiency indices reported in this study are within the bounds of those found in other studies on efficiency of cotton farms. Finally, based on our most preferable model (Model 4), average technical efficiency is about 0.457. Using farm-level data from four counties in west Texas, Chakraborty et al. (2002) estimated average efficiency at 80%. Another study on cotton production found average technical efficiency of 79% for farms in the Çukurova region in Turkey (Gul et al. 2009).

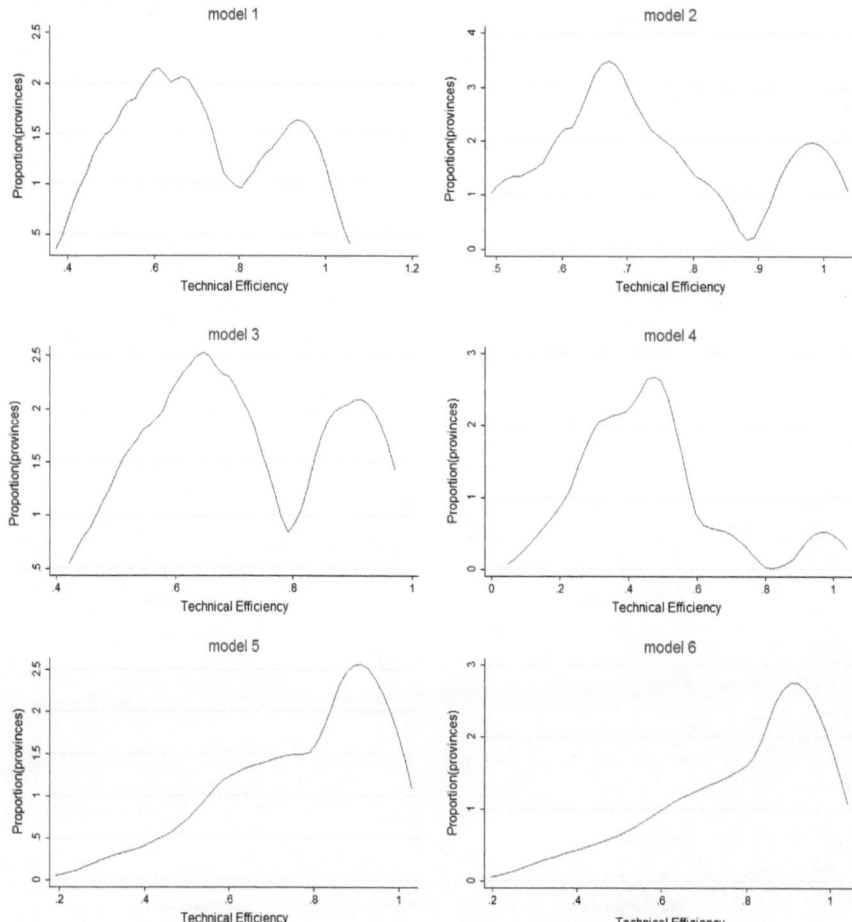

Fig. 5.4 a Technical efficiency distributions of cotton producing provinces for Models 1–6, **b** Technical efficiency distributions of cotton producing provinces for Models 7–12

Another result which can be attained from Table 5.4 is that most of the time-variant models produced larger efficiency levels as compared to time-invariant models.

Figures 5.4 presents the kernel density distribution of technical efficiency estimates for Models 1–12. As an example, consider Models 5 and 6, in which scores for distribution of technical efficiency ranged from 0.25 to 0.97 and 0.25 to 0.98 respectively. According to these results, distribution of technical efficiency in Models 5 to 7 and also in Model 10 had almost the same pattern.

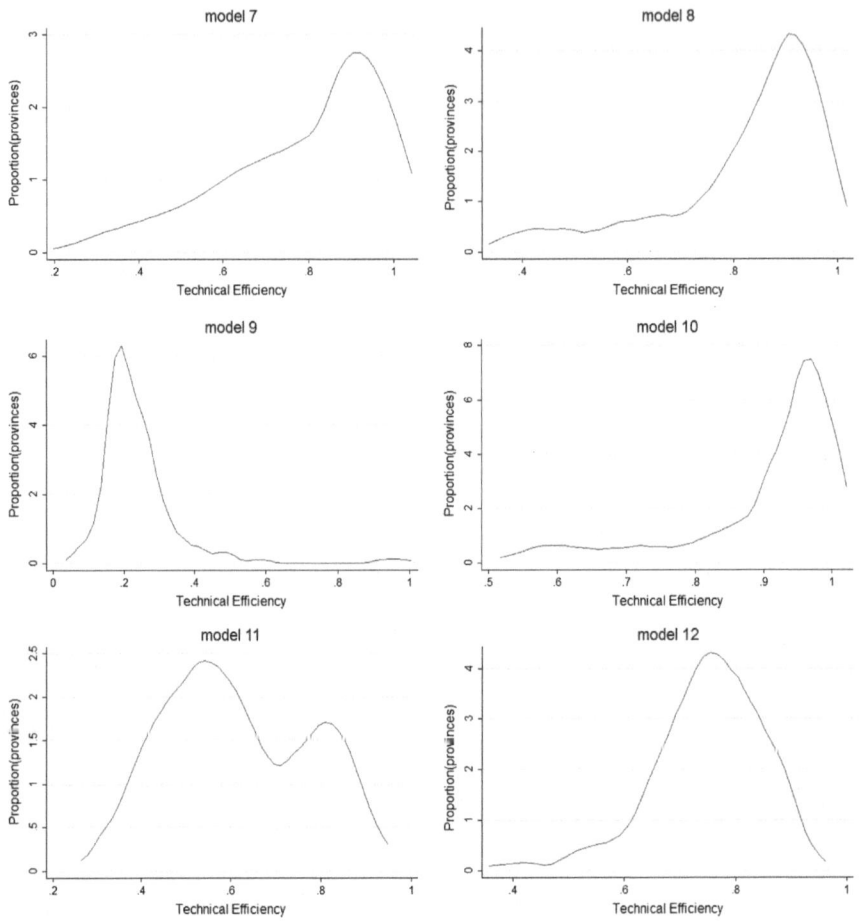

Fig. 5.4 (continued)

5.2.2.3   Technical Efficiency Measurement by Provinces

Descriptive statistics for technical efficiency measured by provinces are presented in Table 5.5. Estimated efficiency scores for different models show that there were differences between provinces in terms of efficiency. For example, estimated technical efficiency according to Model 1 in Khorasan and Qom provinces was 1.00 and 0.50 respectively. According to most of the models reported in Table 5.5, East Azerbaijan, Khorasan and Ardabil were the most efficient provinces in cotton production in Iran. Technical efficiency was estimated to be the lowest in Yazd, Qom and Mazandaran. The results show that according to models in the first group, seven of the 13 provinces, on average (53%) produced below 70% of their potential output due to high technical inefficiencies. Further, the average technical efficiency measure suggests that cotton producing provinces in Iran could increase their production by about 30% through better use of inputs and productivity enhancing extension services. Models 6 and 7 gave a technical efficiency of 0.93 for Tehran province. These results are consistent with Rafati et al. (2011) who concluded that cotton producers in Tehran had a technical efficiency of 93%.

For investigating the performance of different sample provinces and their position as compared to the province with the best practiced technology, we ranked the provinces by the level of their efficiency. Based on Tables 5.5 and 5.6 gives the ranking of different provinces. According to this table Models 1, 3, 5, 6, 7, 8 and 11 had an almost similar ranking for the provinces. According to these models, Khorasan, Ardabil and East Azerbaijan were the most technically efficient provinces and were ranked first to third respectively. Similarly, these models (58% of the models) ranked Yazd, Qom and Mazandaran as the least technically efficient cotton producing provinces in Iran. Model 9 produced the smallest technical efficiency scores for all the provinces.

5.2.2.4   Technical Efficiency Measurement by Years

The yearly mean of provincial technical efficiency for the 12 models is presented in Table 5.7. The results indicate that according to all the time-variant models, except for Model 9, technical efficiency decreased during the study period. A steady decrease in the technical efficiency of Models 6 and 7 verifies the results given in Table 5.2. According to most of the models, 2009 was the most efficient year and 2012 the least efficient year during the study period.

Table 5.8 shows the ranking of different years in the study period. Based on Models 6, 7, 8 and 10, 2000 and 2012 were the most efficient and least efficient years respectively. According to most of the models, there was a decreasing trend in technical efficiency during 2000–2012.

Figure 5.5 illustrates the lower quartile, median and upper quartile efficiencies over time across the 12 models in which the spread of efficiency can be seen. According to Fig. 5.5, Model 4 had the widest efficiency spread and Model 12 had the narrowest spread, slightly narrower than Model 9. Figure 5.5 also shows that the time-series efficiency patterns for Models 5, 6 and 7 were similar.

Pairwise rank-order correlations of Models 1 through 12 are reported in Table 5.9. The correlation between Models 6 and 7 ($\tau = 1.00$), and also between Models 2 and 3 ($\tau = 0.76$) are high. These models seem to be the most consistent in generating similar results, while the results of technical efficiency estimates between Model 10 and Models 1–6 and also between Model 8 and Models 1–7 are to a large extent independent with a low rank-order correlation of less than 0.2.

Table 5.5 Descriptive statistics for technical efficiency measures by provinces

Province	Model 1	Model 2	Model 3	Model 4	Model 5	Model 6	Model 7	Model 8	Model 9	Model 10	Model 11	Model 12
Markazi	0.572	0.627	0.601	0.326	0.660	0.681	0.682	0.931	0.259	0.983	0.496	0.699
Mazandaran	0.520	0.656	0.571	0.249	0.572	0.585	0.585	0.825	0.291	0.939	0.449	0.717
East Azerbaijan	0.917	1.000	0.922	0.486	0.948	0.957	0.957	0.908	0.253	0.974	0.796	0.862
Fars	0.708	0.720	0.716	0.510	0.894	0.909	0.908	0.938	0.203	0.980	0.614	0.760
Kerman	0.641	0.703	0.669	0.343	0.741	0.770	0.770	0.813	0.245	0.938	0.554	0.741
Khorasan	1.000	0.786	0.897	0.972	0.958	0.964	0.964	0.808	0.180	0.890	0.870	0.785
Isfahan	0.627	0.649	0.640	0.404	0.686	0.716	0.713	0.694	0.224	0.776	0.540	0.709
Semnan	0.768	0.809	0.779	0.463	0.881	0.888	0.888	0.842	0.225	0.950	0.664	0.790
Yazd	0.431	0.535	0.471	0.223	0.464	0.498	0.494	0.737	0.267	0.949	0.370	0.638
Tehran	0.894	0.976	0.902	0.485	0.916	0.939	0.933	0.716	0.270	0.700	0.773	0.849
Golestan	0.681	0.671	0.684	0.543	0.771	0.806	0.806	0.924	0.215	0.978	0.592	0.729
Ardabil	0.936	0.968	0.917	0.605	0.957	0.961	0.962	0.880	0.245	0.902	0.813	0.854
Qom	0.504	0.531	0.518	0.327	0.544	0.549	0.545	0.665	0.220	0.767	0.431	0.636

Table 5.6 Ranking of different cotton producing provinces by parametric models

Rank	Model 1	Model 2	Model 3	Model 4	Model 5	Model 6	Model 7	Model 8	Model 9	Model 10	Model 11	Model 12
1	Khorasan	East Azerbaijan	Khorasan	Markazi	Khorasan	Khorasan	Khorasan	Khorasan	Khorasan	East Azerbaijan	Khorasan	East Azerbaijan
2	Ardabil	Ardabil	Ardabil	Fars	Ardabil	Ardabil	Ardabil	Ardabil	Ardabil	Ardabil	Ardabil	Tehran
3	East Azerbaijan	Tehran	East Azerbaijan	Golestan	East Azerbaijan	East Azerbaijan	East Azerbaijan	East Azerbaijan	Golestan	Tehran	East Azerbaijan	Ardabil
4	Tehran	Semnan	Tehran	East Azerbaijan	Tehran	Tehran	Tehran	Tehran	Fars	Khorasan	Tehran	Semnan
5	Semnan	Khorasan	Semnan	Semnan	Fars	Fars	Fars	Fars	East Azerbaijan	Semnan	Semnan	Khorasan
6	Fars	Fars	Fars	Yazd	Semnan	Semnan	Semnan	Semnan	Tehran	Fars	Fars	Fars
7	Golestan	Kerman	Golestan	Mazandaran	Golestan	Golestan	Golestan	Golestan	Semnan	Golestan	Golestan	Kerman
8	Kerman	Golestan	Kerman	Kerman	Kerman	Kerman	Kerman	Kerman	Isfahan	Kerman	Kerman	Golestan
9	Isfahan	Mazandaran	Isfahan	Ardabil	Isfahan	Isfahan	Isfahan	Isfahan	Kerman	Isfahan	Isfahan	Mazandaran
10	Markazi	Isfahan	Markazi	Khorasan	Markazi	Markazi	Markazi	Markazi	Qom	Markazi	Markazi	Isfahan
11	Mazandaran	Markazi	Mazandaran	Isfahan	Mazandaran	Mazandaran	Mazandaran	Mazandaran	Markazi	Mazandaran	Mazandaran	Markazi
12	Qom	Yazd	Qom	Qom	Qom	Qom	Qom	Qom	Mazandaran	Qom	Qom	Yazd
13	Yazd	Qom	Yazd	Tehran	Yazd	Yazd	Yazd	Yazd	Yazd	Yazd	Yazd	Qom

Table 5.7 Descriptive statistics for technical efficiency measures by years

Year	Model 1	Model 2	Model 3	Model 4	Model 5	Model 6	Model 7	Model 8	Model 9	Model 10	Model 11	Model 12
2000	0.708	0.741	0.714	0.489	0.802	0.885	0.885	0.887	0.163	0.951	0.618	0.760
2001	0.708	0.741	0.714	0.495	0.820	0.873	0.873	0.861	0.178	0.933	0.613	0.755
2002	0.708	0.741	0.714	0.498	0.831	0.860	0.860	0.824	0.178	0.898	0.610	0.753
2003	0.708	0.741	0.714	0.498	0.835	0.845	0.845	0.879	0.182	0.923	0.618	0.772
2004	0.708	0.741	0.714	0.494	0.832	0.829	0.829	0.885	0.177	0.911	0.628	0.778
2005	0.708	0.741	0.714	0.487	0.821	0.812	0.812	0.815	0.197	0.872	0.619	0.768
2006	0.708	0.741	0.714	0.477	0.803	0.794	0.794	0.837	0.245	0.894	0.611	0.745
2007	0.708	0.741	0.714	0.463	0.778	0.774	0.774	0.842	0.235	0.921	0.614	0.763
2008	0.708	0.741	0.714	0.447	0.749	0.753	0.753	0.838	0.264	0.889	0.625	0.771
2009	0.708	0.741	0.714	0.428	0.717	0.731	0.731	0.876	0.345	0.931	0.633	0.784
2010	0.708	0.741	0.714	0.408	0.688	0.708	0.708	0.671	0.295	0.870	0.581	0.695
2011	0.708	0.741	0.714	0.387	0.665	0.684	0.684	0.765	0.341	0.867	0.609	0.739
2012	0.708	0.741	0.714	0.366	0.650	0.658	0.658	0.701	0.297	0.863	0.581	0.685

Table 5.8 Ranking of different years

Rank	Model 4	Model 5	Model 6	Model 7	Model 8	Model 9	Model 10	Model 11	Model 12
1	2002	2003	2000	2000	2000	2009	2000	2009	2009
2	2003	2004	2001	2001	2004	2011	2001	2004	2004
3	2001	2002	2002	2002	2003	2012	2009	2008	2003
4	2004	2005	2003	2003	2009	2010	2003	2005	2008
5	2000	2001	2004	2004	2001	2008	2007	2003	2005
6	2005	2006	2005	2005	2007	2006	2004	2000	2007
7	2006	2000	2006	2006	2008	2007	2002	2007	2000
8	2007	2007	2007	2007	2006	2005	2006	2001	2001
9	2008	2008	2008	2008	2002	2003	2008	2006	2002
10	2009	2009	2009	2009	2005	2002	2005	2002	2006
11	2010	2010	2010	2010	2011	2001	2010	2011	2011
12	2011	2011	2011	2011	2012	2004	2011	2010	2010
13	2012	2012	2012	2012	2010	2000	2012	2012	2012

Table 5.9 Kendall's rank-order correlation between the technical efficiency of Models 1–12

	Model 1	Model 2	Model 3	Model 4	Model 5	Model 6	Model 7	Model 8	Model 9	Model 10	Model 11
Model 2	0.69[a]										
Model 3	0.85[a]	0.76[a]									
Model 4	0.61[a]	0.42[a]	0.55[a]								
Model 5	0.76[a]	0.63[a]	0.74[a]	0.66[a]							
Model 6	0.73[a]	0.63[a]	0.72[a]	0.64[a]	0.89[a]						
Model 7	0.73[a]	0.63[a]	0.72[a]	0.64[a]	0.89[a]	1.00[a]					
Model 8	0.09[a]	0.09[a]	0.13[a]	0.20[a]	0.16[a]	0.16[a]	0.16[a]				
Model 9	−0.10[a]	−0.03	−0.07	−0.24[a]	−0.23[a]	−0.25[a]	−0.25[a]	−0.13[a]			
Model 10	−0.01	0.03	0.04	0.02	0.04	0.06	0.06[a]	0.56[a]	−0.11[a]		
Model 11	0.86[a]	0.67[a]	0.81[a]	0.65[a]	0.76[a]	0.73[a]	0.73[a]	0.15[a]	−0.09[a]	−0.02	
Model 12	0.49[a]	0.56[a]	0.53[a]	0.41	0.51[a]	0.50[a]	0.50[a]	0.24[a]	0.01	−0.05	0.61[a]

Notes: Significant at less than 1% ([a])

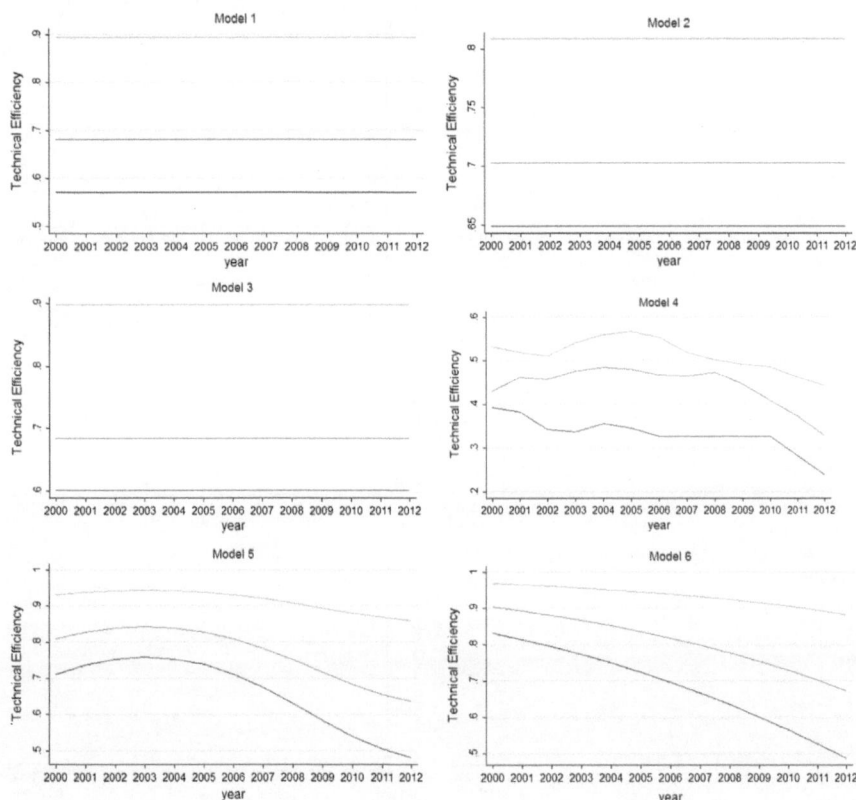

Fig. 5.5 a The mean, first and third quartile values of technical efficiency of different cotton producing provinces for Model 1–6, **b** The mean, first and third quartile values of technical efficiency of different cotton producing provinces for Model 7–12

Table 5.10 shows Kendall's rank-order correlation for the persistent technical efficiency measure between Models 11 and 12. According to this table, assessments of residual technical efficiency for these two models were to a large extent positively correlated with a rank-order correlation of 0.69. The results indicate that persistent and residual technical efficiencies were independent.

In Fig. 5.6 the scatter plot matrices for Models 1–12 graphically illustrate the differences between the models in the ranking of provinces. The straight lines in the graph indicate a perfect match between two compared models. These results prove the findings given in Table 5.9.

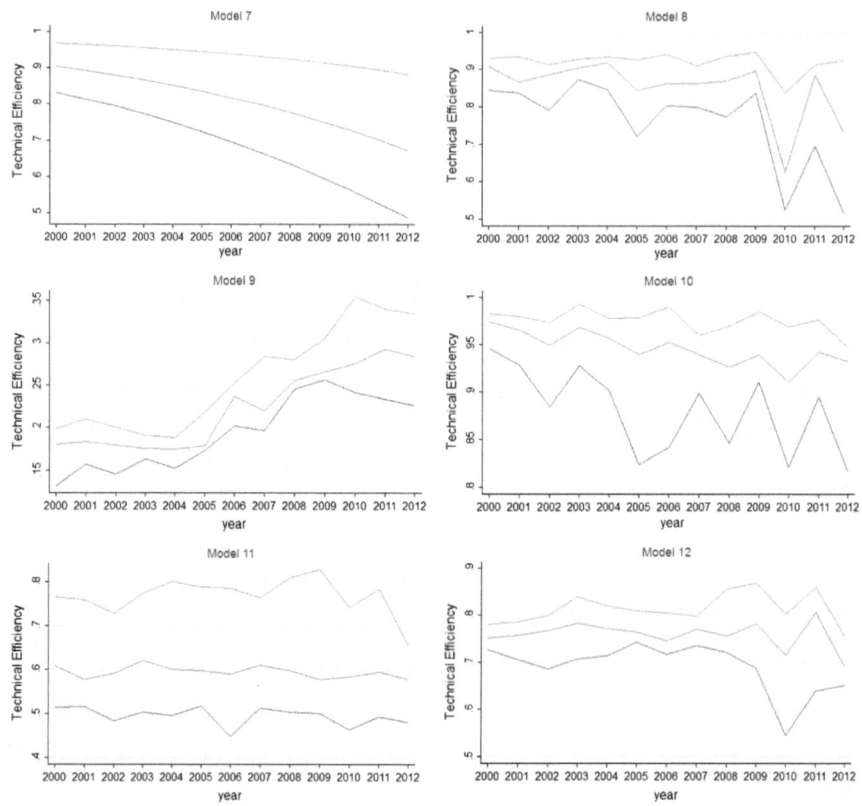

Fig. 5.5 (continued)

Table 5.10 Kendall's rank-order correlation between the technical efficiency of Models 11 and 12 for persistent technical efficiency (PTE) and residual technical efficiency (RTE) estimates

	Model 11 PTE	Model 11 RTE	Model 12 PTE
Model 11 RTE	−0.0280		
Model 12 PTE	0.6905[a]	−0.0331	
Model 12 RTE	0.3470[a]	0.5314[a]	0.3957[a]

Notes: Significant at less than 1% ([a])

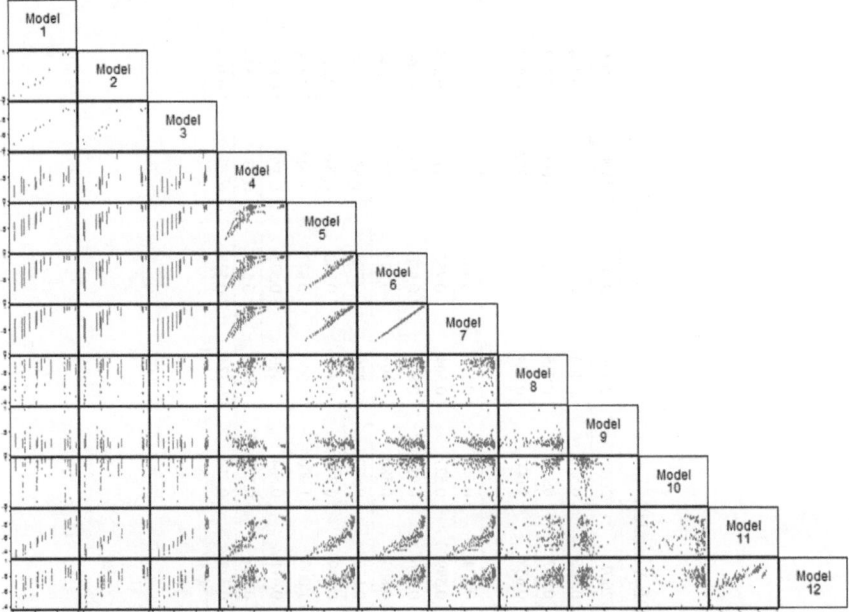

Fig. 5.6 Scatter plot matrices of pairwise technical efficiency estimates for Models 1–12

5.2.3 Non-parametric Models

5.2.3.1 Provincial Technical Efficiency Measurement

We ran two full frontier models (DEA and FDH) and two partial frontier models (Order-α and Order-m with different values for α and m) using a Window analysis. Tables 5.11, 5.12, 5.13 and 5.14 give the efficiency scores obtained using Window-DEA, Window-FDH, Window Order-m and Window Order-α models for different cotton producing provinces. To conserve space only the average technical efficiency of provinces in different years of each Window model are presented. According to the Window-DEA model in Table 5.11, the most efficient cotton producing provinces were Khorassan, East Azerbaijan, Tehran and Ardebil while the least efficient provinces were Mazandaran and Qom.

As an example consider Khorasan province which saw a decreasing trend in technical efficiency during the study period except in 2004, 2008 and 2010. Based on this model, East Azerbaijan experienced dramatic fluctuations in technical efficiency during the study period.

Table 5.11 Mean technical efficiency of different windows during 2000–2012 (Window-DEA model)

	2000	2001	2002	2003	2004	2005	2006	2007	2008	2009	2010	2011	2012
Markazi	0.852	0.762	0.587	0.855	0.497	0.815	0.357	0.626	0.559	0.403	0.438	0.495	0.404
Mazandaran	0.295	0.286	0.335	0.375	0.334	0.236	0.234	0.362	0.152	0.085	0.019	0.009	0.272
East Azerbaijan	1.000	0.906	0.991	0.899	1.000	1.000	0.932	0.658	1.000	0.854	0.501	0.732	0.962
Fars	1.000	0.940	0.834	1.000	0.707	1.000	0.865	0.874	0.564	0.511	0.465	0.667	0.678
Kerman	0.754	0.602	0.511	0.637	0.895	0.972	0.909	0.599	0.52·	0.564	0.757	0.778	0.365
Khorasan	1.000	0.985	0.977	0.965	0.987	0.956	0.927	0.906	0.911	0.893	0.936	0.966	0.966
Isfahan	0.700	0.623	0.551	1.000	0.733	0.519	0.531	0.803	0.589	0.494	0.252	0.247	0.344
Semnan	0.863	0.756	0.715	0.895	0.876	0.665	0.564	0.522	0.937	0.876	0.704	0.839	0.399
Yazd	1.000	0.617	0.701	0.922	0.693	0.982	1.000	0.449	0.345	0.469	0.280	0.208	0.216
Tehran	1.000	0.996	0.659	0.781	0.943	0.964	0.908	1.000	1.000	1.000	0.677	1.000	0.398
Golestan	0.655	0.903	0.681	1.000	1.000	0.920	0.615	0.611	0.564	0.480	0.534	0.328	0.856
Ardabil	0.922	0.764	0.942	0.761	0.987	0.753	0.936	0.626	0.722	1.000	0.579	0.659	0.572
Qom	0.606	0.977	1.000	0.681	0.649	0.251	0.313	0.333	0.417	0.324	0.225	0.256	0.103

Table 5.12 Mean technical efficiency of different windows during 2000–2012 (Window-FDH model)

	2000	2001	2002	2003	2004	2005	2006	2007	2008	2009	2010	2011	2012
Markazi	1.000	1.000	0.826	1.000	0.871	1.000	0.983	1.000	1.000	0.469	0.895	1.000	1.000
Mazandaran	0.600	0.600	0.600	0.600	0.600	0.600	0.600	0.600	0.600	0.600	0.600	0.600	0.600
East Azerbaijan	1.000	1.000		1.000	1.000	1.000	1.000	0.945	1.000	1.000	1.000	1.000	1.000
Fars	1.000	1.000	1.000	1.000	1.000	1.000	1.000	1.000	1.000	1.000	1.000	1.000	1.000
Kerman	1.000	1.000	1.000	1.000	1.000	1.000	1.000	1.000	1.000	1.000	1.000	1.000	0.897
Khorasan	1.000	1.000	1.000	1.000	1.000	1.000	1.000	1.000	1.000	1.000	1.000	1.000	1.000
Isfahan	0.984	0.832	0.957	1.000	1.000	0.997	0.745	1.000	1.000	1.000	0.298	0.438	0.535
Semnan	1.000	1.000	1.000	1.000	1.000	0.991	0.905	1.000	1.000	1.000	1.000	1.000	1.000
Yazd	1.000	1.000	1.000	1.000	1.000	1.000	1.000	0.985	1.000	1.000	1.000	1.000	0.982
Tehran	1.000	1.000	0.924	1.000	1.000	1.000	0.961	1.000	1.000	1.000	1.000	1.000	1.000
Golestan	1.000	1.000	0.998	1.000	1.000	1.000	1.000	1.000	1.000	1.000	1.000	0.771	1.000
Ardabil	1.000	1.000	1.000	1.000	1.000	1.000	1.000	1.000	1.000	1.000	1.000	1.000	1.000
Qom	0.948	1.000	1.000	1.000	0.855	0.380	0.494	0.594	0.520	0.980	0.386	0.500	0.112

Table 5.13 Mean technical efficiency of different windows during 2000–2012 (Window Order-m model)

	2000	2001	2002	2003	2004	2005	2006	2007	2008	2009	2010	2011	2012
Markazi	1.014	1.028	0.837	1.367	0.878	1.003	1.030	1.026	1.197	0.497	0.901	1.142	1.018
Mazandaran	0.298	0.257	0.269	0.269	0.342	0.333	0.685	0.487	0.478	0.641	0.275	0.425	0.956
East Azerbaijan	1.040	1.068	1.034	1.087	1.076	1.005	1.062	0.898	1.043	1.020	1.083	1.000	1.109
Fars	1.000	1.000	1.000	1.000	1.000	1.000	1.000	1.000	1.000	1.000	1.000	1.000	1.000
Kerman	0.648	0.595	0.612	0.519	0.683	1.012	0.781	0.947	0.713	0.845	0.974	1.008	0.715
Khorasan	1.394	1.394	1.386	1.394	1.394	1.394	1.394	1.394	1.394	1.386	1.360	1.394	1.394
Isfahan	0.984	0.661	0.933	1.031	1.002	0.997	0.748	1.000	1.001	0.805	0.272	0.327	0.384
Semnan	1.001	1.003	1.005	1.002	0.813	0.954	0.906	1.008	1.006	1.006	1.002	1.019	0.625
Yazd	0.845	1.030	0.922	0.968	0.836	1.068	0.925	0.951	1.014	0.861	0.771	1.049	0.831
Tehran	1.000	0.866	0.675	1.013	1.063	1.071	0.873	1.025	1.013	1.273	0.920	1.340	0.696
Golestan	1.000	1.000	0.998	1.000	1.000	1.000	1.002	1.000	1.000	1.001	0.897	0.771	1.000
Ardabil	1.000	1.000	1.000	1.000	1.000	1.001	1.011	0.765	0.965	1.013	1.000	1.007	1.005
Qom	0.948	1.000	1.011	1.003	0.860	0.392	0.470	0.480	0.528	0.606	0.324	0.375	0.113

Table 5.14 Mean technical efficiency of different windows during 2000–2012 (Window order-α model)

	2000	2001	2002	2003	2004	2005	2006	2007	2008	2009	2010	2011	2012
Markazi	0.760	0.952	0.580	0.964	0.574	0.687	1.002	0.843	1.025	0.610	0.654	0.969	0.236
Mazandaran	0.611	0.132	0.245	0.352	0.776	0.663	1.076	1.212	1.254	1.689	0.819	1.798	1.376
East Azerbaijan	1.726	1.535	1.814	1.952	3.056	0.318	2.119	1.132	1.587	1.298	1.821	1.001	1.206
Fars	1.000	1.000	1.000	1.000	1.000	1.000	1.000	1.000	1.000	1.000	1.000	1.000	1.000
Kerman	1.156	0.918	0.788	0.866	1.000	1.156	0.984	1.401	0.862	0.881	1.383	1.157	0.715
Khorasan	1.400	1.390	0.992	1.000	1.000	1.000	1.000	1.000	1.000	0.992	0.993	1.000	1.000
Isfahan	1.000	0.861	0.957	2.178	1.000	0.997	0.873	1.000	1.000	0.951	0.389	1.380	0.550
Semnan	1.098	1.260	1.226	1.282	1.404	0.995	0.976	1.000	1.310	1.288	1.106	1.000	0.623
Yazd	1.000	1.290	1.130	1.306	1.077	1.931	1.088	1.237	1.329	1.253	1.134	1.500	1.000
Tehran	1.000	1.272	0.934	1.426	1.514	1.615	1.732	1.101	1.931	1.259	2.458	1.382	4.000
Golestan	1.000	1.000	0.998	1.000	1.000	1.000	1.000	1.000	1.000	1.000	0.931	0.771	1.000
Ardabil	1.000	1.000	1.000	1.000	1.000	1.000	1.678	1.007	0.982	1.451	1.000	1.934	1.070
Qom	1.000	1.000	1.000	1.326	1.072	0.606	0.680	0.666	0.654	0.913	0.431	0.374	0.242

Table 5.15 Descriptive statistics for average efficiency scores by non-parametric models

Intervals	Model 13 (Window DEA model)	Model 14 (Window FDH model)	Model 15 (Window order-m model)	Model 16 (Window Order-α model)
=>1.00	0.000	0.550	0.390	0.690
0.90–1.00	0.300	0.250	0.300	0.120
0.80–0.90	0.230	0.040	0.100	0.030
0.70–0.80	0.170	0.000	0.040	0.030
0.60–0.70	0.100	0.110	0.020	0.070
0.50–0.60	0.100	0.020	0.050	0.020
0.40–0.50	0.050	0.006	0.060	0.010
0.30–0.40	0.110	–	0.010	0.001
0.20–0.30	0.020	–	0.010	–
0.10–0.20	0.020	–	–	–
0.00–0.10	0.006	–	–	–
Mean	0.680	0.720	0.910	1.110
Skewness	−0.770	−1.180	−0.560	1.330
Kurtosis	2.700	4.660	3.680	6.480

With respect to Window-FDH Model, Table 5.12 shows that Khorasan, Fars and Ardabil provinces were fully efficient during 2000–2012. This model also reports that the technical efficiency of most of the provinces in different years was about 100%. Like the Window-DEA model, the Window-FDH model also classifies Mazandaran and Qom as the least efficient cotton producing provinces in Iran.

Table 5.13 presents mean technical efficiency scores of cotton producing provinces based on the Window Order-m model. According to this model, mean technical efficiency of Khorasan province was more than 100% during the study period. Therefore, this province is classified as a super-efficient province. East Azerbaijan also had technical efficiency scores that were more than 100% during 2000–2012 (except for 2007). Like the full frontier models, this partial frontier model classifies Mazandran and Qom as the least efficient provinces.

Results of mean technical efficiency measurement based on Window Order-α are presented in Table 5.14. According to this table and as expected, the mean technical efficiency of cotton producing provinces during the study period was even larger than the other models. This model classifies more provinces as super- efficient. Table 5.15 gives a comparison of the different non-parametric models used in the study.

5.2.3.2   Distribution of Technical Efficiency

Table 5.15 shows some of the distributional characteristics of the efficiency scores of provinces using Window-DEA, Window-FDH, Window Order-m and Window Order-α models. It also presents the percentage of efficient and super-efficient provinces. According to the Window-DEA model, a majority of the provinces (about 30%) were in the efficiency range of 90–100%. In this model mean technical efficiency was about 68; 53% of the provinces had more than 80% technical efficiency. Mean technical efficiency according to the Window-FDH model was 72%. Most of the provinces were in the range of equal to or larger than 1.00 efficiency. When we consider the FDH model, 55% of the provinces were efficient or super-efficient. This number is significantly larger than the 30% efficient provinces obtained using Window-DEA. There are two main reasons for the differences between FDH and DEA: (i) The convergence rate in the FDH model is slow, and (ii) Assumption of convexity in the DEA model. Other studies which have used DEA and FDH models for the same dataset have also reported higher technical efficiency scores for the FDH model as compared to the DEA model (see De Borger et al. 1994; De Witte and Marques 2010; Gabdo et al. 2014).

Columns 4 and 5 of Table 5.15 give efficiency estimates of the Window Order-m (Model 15) and Window Order-α (Model 16). Mean technical efficiency measurement according to Models 15 and 16 was 91 and 111% respectively. According to these models, 39 and 69% of the provinces were efficient or super-efficient. Silva et al. (2016) study also reported that some of the firms were super-efficient with efficiency scores larger than 1.

When we drop the convexity assumption (that is, move from Window-DEA to Window-FDH), the estimated efficiency scores become higher (this is as expected since the best practice frontier then wraps itself closer around the data). In a similar way, estimated technical efficiency scores become even higher when we use partial frontier approaches (Order-m and Order-α).

5.2.3.3 Ranking of Different Cotton Producing Provinces

The most interesting results of efficiency measurement in such cases are those provided by the complete listing of the efficiency scores for each province. Table 5.16 presents provinces' ranks determined by different non-parametric models. According to full frontier models (DEA and FDH), Khorasan was the most efficient province. East Azerbaijan and Tehran ranked second and third in technical efficiency scores while Mazandaran and Qom were the least efficient provinces. Partial frontier models ranked Khoarasan and East Azerbaijan provinces as the most efficient cotton producing provinces in the country. According to this table, although there were differences in the exact level of efficiency depending on the approach used, efficiency rankings from various approaches tend to support similar conclusions about the relative performance of provinces.

Table 5.16   Scores and ranking values obtained by non-parametric models

Provinces	Model 13 Window DEA scores	Model 14 Window FDH scores	Model 15 Window order-m scores	Model 16 Window order-α scores	Model 13 Window-DEA ranking	Model 14 Window FDH ranking	Model 15 Window order-m ranking	Model 16 Window order-α ranking
Markazi	0.589	0.926	0.995	0.749	10	8	4	12
Mazandaran	0.241	0.600	0.438	0.625	13	11	13	13
East Azerbaijan	0.880	0.999	1.041	1.735	2	2	2	1
Fars	0.777	1.000	1.000	1.000	5	1	3	8
Kerman	0.668	0.992	0.775	1.020	8	4	11	7
Khorasan	0.961	1.000	1.396	1.396	1	1	1	2
Isfahan	0.568	0.830	0.780	0.933	11	9	10	10
Semnan	0.739	0.992	0.950	1.120	6	5	8	6
Yazd	0.606	0.997	0.929	1.251	9	6	9	3
Tehran	0.871	0.991	0.987	0.241	3	3	5	4
Golestan	0.704	0.982	0.975	0.976	7	7	7	9
Ardabil	0.786	1.000	0.982	1.163	4	1	6	5
Qom	0.472	0.657	0.624	0.766	12	10	12	11
Mean	0.682	0.923	0.913	1.075	–	–	–	–

5.2.3.4   Kendall's Rank-Order Correlation Between Non-parametric Models

We calculated the Kendall rank-order correlation coefficients (r) to determine how close the implied rankings of provinces were in each of the models. Coefficient r is essentially a measure of association derived from ranks of observations between two series. A value of $r = 1$ (or $r = -1$) indicates a perfectly positive (negative) rank-order correlation, while $r = 0$ indicates that no correlation exists between the efficiency of the two provinces. Pairwise rank-order correlations of different models are reported in Table 5.17. The main results of this table can be summed as: (i) The correlation between Window-DEA and Window-FDH was relatively high at 0.84. This implies that in our study Window-DEA and Window-FDH ranked provinces in almost the same order, (ii) Rank-order correlation between other models was also positive and significant (except for the correlation between Models 13 and 16) and (iii) Rank-order correlation between two partial frontier models (Window Order-m and Window Order-α) was high and significant ($\tau = 0.49$).

5.3   Results of Sugar Beet Production

5.3.1   Descriptive Results

In this section we present the results of technical efficiency modeling for sugar beet production in Iran. The results are based on information about inputs and outputs

Table 5.17   Kendall's rank-order correlation between the technical efficiency of non-parametric models

	Model 13 (Window- DEA)	Model 14 (Window- FDH)	Model 15 (Window order-m)	Model 16 (Window order-α)
Model 13 (Window DEA)	1.00			
Model 14 (Window FDH)	0.84[a]	1.00		
Model 15 (Window Order-m)	0.55[a]	0.45[a]	1.00	
Model 16 (Window Order-α)	0.43	0.26[a]	0.49[a]	1.00

Notes: Significant at less than 1% ([a])

from 11 sugar beet producing provinces during 2000–2012 (Markazi, West Azerbaijan, Kermanshah, Fars, Kerman, Khorasan, Isfahan, Chaharmahal and Bakhtiari, Lorestan, Semnan and Qazvin provinces). Statistics on harvested area, production and yield of sugar beet crop are shown in Figs. 5.7, 5.8 and 4.9 respectively. Based on Figs. 5.7 and 5.8, Khorasan and West Azerbaijan provinces had the largest harvested area in the country. These provinces also had the highest amount of sugar beet production in the country. Markazi and Kerman had the minimum harvested area and the minimum production of this crop among the studied provinces. According to Fig. 5.9, West Azerbaijan and Kerman had the highest and the lowest yields of sugar beet production respectively among the provinces.

Descriptive statistics for input and output variables used in different parametric and non-parametric models for sugar beet crop are presented in Table 5.18. Sugar beet yields in the country were about 32,206 kg/ha, with the highest and lowest yields in West Azerbaijan (42,539 kg/ha) and Kerman (26,162 kg/ha) provinces. After West Azerbaijan, Kermanshah and Qazvin provinces had the highest yields in sugar beet production.

The highest and lowest amount of pesticides were used in Qazvin (7.82 kg/ha) and Kermanshah (2.98 kg/ha) respectively. In Kermanshah, the amount of pesticides used varied from 0.90 to 5.01 kg/ha. The average labor use was about 79 persons per hectare. Semnan province had the highest labor use for sugar beet production. Chemical fertilizers are another input that is used in the modeling of sugar beet production. The average consumption of this input was 643 kg/ha. Kerman, Isfahan, Chaharmahal, Bakhtiari and Semnan provinces had the highest chemical fertilizer use among the provinces. According to the second part of Table 4.18, the average use of machinery amongst the sugar beet-producing provinces was about 86%.

Table 5.18 Summary statistics of input and output data for the sugar beet crop (2000–2012), 11 provinces and 13 years, 143 observations

Variable	Province	N	Mean	Std. dev.	Minimum	Maximum
Production function variables:						
y output (Kilograms)	Markazi	13	31178.72	8427.56	21454.78	47412.82
	West Azerbaijan	13	42539.16	6479.25	30110.32	53070.01
	Kermanshah	13	35711.15	11455.90	19603.44	64231.00
	Fars	13	28789.07	4495.12	20925.78	35395.15
	Kerman	13	26162.98	4403.29	18775.36	35002.10
	Khorasan	13	30993.37	1944.82	26460.81	33500.91
	Isfahan	13	30995.53	5531.41	21227.35	42381.28
	Chaharmahal and Bakhtiari	13	28756.12	3988.68	22548.13	34498.38
	Lorestan	13	29934.26	8489.04	18555.00	46921.79
	Semnan	13	33615.18	5380.07	28409.09	48332.79
	Qazvin	13	35601.04	8615.50	24936.21	56492.23
x_p pesticide (Kilograms)	Markazi	13	5.80	2.91	2.80	14.32
	West Azerbaijan	13	5.46	1.46	1.85	7.76
	Kermanshah	13	2.98	1.32	0.90	5.01
	Fars	13	5.83	1.39	3.14	8.27
	Kerman	13	4.20	3.28	0.72	11.54
	Khorasan	13	3.03	1.60	0.88	6.11
	Isfahan	13	4.37	1.05	2.64	6.10
	Chaharmahal and Bakhtiari	13	5.15	2.90	2.15	11.46
	Lorestan	13	3.69	1.72	1.00	6.79
	Semnan	13	6.19	3.11	1.91	11.00
	Qazvin	13	7.82	4.50	3.49	15.89
x_l labor (man-day)	Markazi	13	88.32	12.46	62.00	106.60
	West Azerbaijan	13	84.71	20.84	50.35	130.95
	Kermanshah	13	65.08	27.36	25.14	108.54
	Fars	13	84.93	20.02	35.33	106.87
	Kerman	13	63.74	18.38	33.98	97.51
	Khorasan	13	81.85	18.44	41.98	115.91
	Isfahan	13	88.27	25.42	38.72	139.76
	Chaharmahal and Bakhtiari	13	80.00	16.82	50.38	113.83
	Lorestan	13	71.31	24.64	31.34	129.67
	Semnan	13	92.85	22.92	55.54	139.54
	Qazvin	13	66.41	19.43	42.03	101.56

(continued)

Table 5.18   (continued)

Variable	Province	N	Mean	Std. dev.	Minimum	Maximum
x_{cf} chemical fertilizers (Kilograms)	Markazi	13	572.29	102.92	405.96	769.36
	West Azerbaijan	13	554.27	113.57	372.90	716.76
	Kermanshah	13	674.93	210.46	240.00	972.23
	Fars	13	695.54	102.88	525.61	971.22
	Kerman	13	748.67	179.51	366.95	930.35
	Khorasan	13	512.59	93.37	294.10	701.43
	Isfahan	13	731.26	180.78	456.23	1122.67
	Chaharmahal and Bakhtiari	13	733.05	181.32	339.27	996.25
	Lorestan	13	428.20	87.33	238.40	561.69
	Semnan	13	748.56	222.45	418.93	1180.19
	Qazvin	13	673.89	109.77	513.43	869.89
Inefficiency determinant variables in inefficiency function:						
Z_1 rate of chemical fertilizers per total fertilizers	Markazi	13	1.00	0.00	0.99	1.00
	West Azerbaijan	13	1.00	0.00	0.99	1.00
	Kermanshah	13	1.00	0.00	1.00	1.00
	Fars	13	1.00	0.00	0.99	1.00
	Kerman	13	0.99	0.00	0.98	0.99
	Khorasan	13	1.00	0.00	0.99	1.00
	Isfahan	13	1.00	0.00	1.00	1.00
	Chaharmahal and Bakhtiari	13	1.00	0.01	0.98	1.00
	Lorestan	13	0.99	0.01	0.98	1.00
	Semnan	13	1.00	0.00	0.99	1.00
	Qazvin	13	1.00	0.00	0.99	1.00
Z_2 machinery (per cent)	Markazi	13	45.71	19.31	22.20	100.00
	West Azerbaijan	13	90.57	7.54	79.87	100.00
	Kermanshah	13	66.18	27.37	8.13	100.00
	Fars	13	53.25	20.16	12.58	84.21
	Kerman	13	9.96	6.47	1.75	25.00
	Khorasan	13	47.86	13.89	24.52	74.56
	Isfahan	13	32.48	23.70	3.42	66.17
	Chaharmahal and Bakhtiari	13	91.13	7.91	77.61	100.00
	Lorestan	13	56.29	28.67	14.95	100.00
	Semnan	13	65.20	29.85	21.49	100.00
	Qazvin	13	88.02	20.75	35.56	100.00

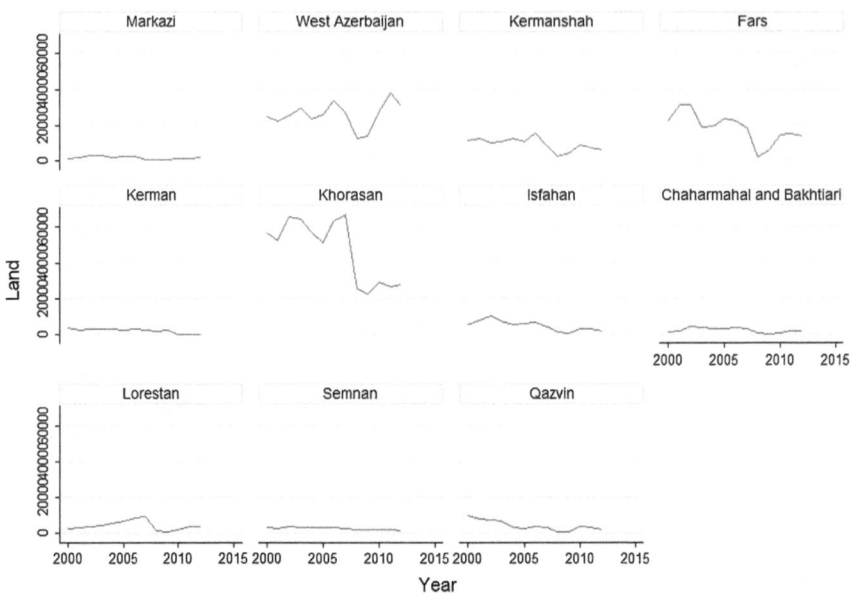

Fig. 5.7 Sugar beet harvested area in different provinces (2000–2012)

5.3.2   Parametric Models

5.3.2.1   Results of Stochastic Frontier Models

In case of sugar beet production, the flexible translog functional form was more consistent with our dataset. Therefore, this functional form was applied to all the 12 models to measure technical efficiency of sugar beet producing provinces in the country. We also normalized input variables at the sample mean. Hence, the first-order coefficients of the input variables are the elasticity of each variable. Here we only discuss the first-order coefficients of the variables. The estimation results for sugar beet production are given in Table 5.19.

According to Table 5.19 and based on Model 1, the estimated output elasticities of pesticides, chemical fertilizers and labor were all positive and statistically significant at the 5% level, indicating that, as expected, production increased with such inputs. The elasticities for pesticides in all the models were in the range of 0.11–0.43 and were statistically significant in all models (except for Models 6 and 7). Elasticity of this input was 0.169 in Model 8. Elasticity of chemical fertilizers ranged from

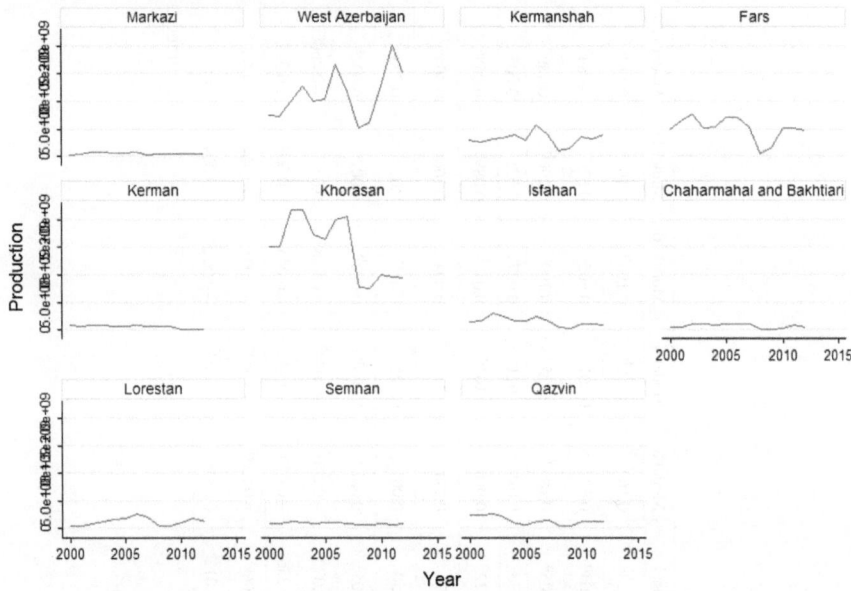

Fig. 5.8 Sugar beet production in different provinces (2000–2012)

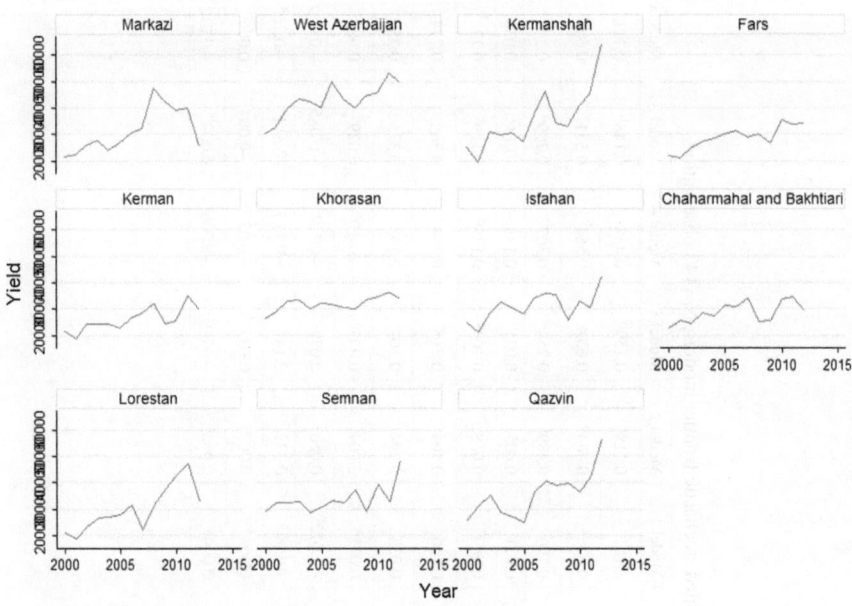

Fig. 5.9 Sugar beet yield in different provinces (2000–2012)

Table 5.19 Estimated stochastic frontier models, N = 143 observation

	Model 1	Model 2	Model 3	Model 4	Model 5	Model 6	Model 7	Model 8	Model 9	Model 10	Model 11	Model 12
x_p	0.161[b]	0.118[b]	0.159[b]	0.438	0.148[b]	0.111[b]	0.113[b]	0.169[b]	0.160[b]	0.195[b]	0.161[c]	0.118[c]
x_{cf}	0.626[a]	0.484[b]	0.602[c]	0.554[b]	0.531[a]	0.604[c]	0.604[b]	0.622[c]	0.601[c]	0.605[c]	0.626[a]	0.484[b]
x_l	0.196[b]	0.489[c]	0.254[c]	0.307[b]	0.299[b]	0.305[b]	0.305[b]	0.189[b]	0.225[b]	0.190[b]	0.196[b]	0.489[b]
time	0.021	0.005	0.021	0.146	-0.141	0.038	0.038	0.019	0.021	0.027	0.021	0.005
$(1/2)x_p^2$	0.089[b]	0.058[b]	0.103[c]	0.138[c]	0.119[c]	0.122[b]	0.122[b]	0.080[c]	0.103[c]	0.047[b]	0.089[c]	0.058[c]
$(1/2)x_{cf}^2$	0.166	0.404	0.215	0.354	0.368	0.274	0.274	0.171	0.215	0.094	0.166	0.404
$(1/2)x_l^2$	0.033	0.294	0.104	0.193	0.325	0.250	0.250	0.001	0.105	-0.037	0.033	0.294
$(1/2)x_t^2$	0.002[a]	0.004[a]	0.003[a]	-0.010[a]	0.029[b]	0.005[a]	0.005[a]	0.002[a]	0.003[b]	0.002[b]	0.002[a]	0.004[a]
$(x_p)*(x_{cf})$	-0.072[b]	0.060[a]	0.079[b]	-0.121[b]	-0.085[b]	-0.079[b]	-0.079[a]	-0.070[b]	-0.078[b]	-0.028[c]	0.072[c]	-0.060[c]
$(x_c)*(x_l)$	-0.093	0.322	0.141	-0.213	-0.314	-0.236	-0.236	-0.079	-0.142	-0.045	-0.093	-0.322
$(x_p)*(x_t)$	0.004[a]	0.009[a]	0.003[a]	-0.027[a]	-0.004[b]	0.006[a]	0.006[a]	-0.005[b]	-0.003[b]	-0.006[b]	-0.004[b]	-0.009[b]
$(x_c)*(x_t)$	0.004[a]	0.002[a]	-0.003[a]	-0.006[a]	0.003[b]	-0.012[b]	-0.012[a]	0.005[a]	-0.003[b]	-0.002[b]	-0.004[b]	0.002[b]

(continued)

Table 5.19 (continued)

		Model 1	Model 2	Model 3	Model 4	Model 5	Model 6	Model 7	Model 8	Model 9	Model 10	Model 11	Model 12
β_0		18.610	18.582	19.181	18.407	21.384	18.806	18.806	18.570	18.571[a]	0.195[c]	18.610	18.582
Gamma	γ_1	–	–	–	–	0.166[a]	–	–	–	–	–	–	–
	γ_2	–	–	–	–	−0.024[b]	–	–	–	–	–	–	–
	γ	–	–	–	–	–	0.094[b]	0.092[b]	–	–	–	–	–
	Time	–	–	–	–	–	–	–	–	–	–	–	–
	Time2	–	–	–	–	–	–	–	–	–	–	–	–
	TimeT	–	–	–	–	–	–	–	–	–	–	–	–
μ		–	–	0.613	–	0.298[a]	0.873[a]	0.280	–	–	–	–	–
σ_u^2	cons	0.289	0.291	−2.942		3.067[a]	1.840[aa]	−4.116[a]	−2.074[a]	−9.544	15.957	0.289	0.291
	Z1	–	–	–		–	–	–	1.140[a]	–	–	–	–
	Z2	–	–	–		–	–	–	−0.053[a]	–	–	–	–
σ_v^2		0.234	0.249	−2.287[a]		−3.141[a]	−3.049[a]	−3.049[a]	−3.144[a]	−2.987	−3.139[a]	0.234	0.249
Returns to scale		0.983	1.091	1.015	1.299	0.978	1.02	1.022	1.61	0.986	0.99	0.983	1.091
R-squared		0.944	0.951		0.978							0.944	0.951
Log-likelihood				−3.873		0.048	−0.580	−0.582[b]	4.952	12.508	15.595		

Notes: Significant at less than 1% ([a]), 1–5% ([b]), and 5–10% ([c]) levels of significance

0.48 to 0.63 and was significant in all the models. The elasticities for labor in all the models were positive and significant and ranged from 0.18 (in Model 8) to 0.48 (in Model 2). According to Model 4, labor elasticity was 0.30, meaning that if labor increased by 1%, sugar beet production will increase by 0.30%. Like the results for the cotton crop, chemical fertilizers had the largest partial elasticity among the inputs. Therefore, it can be concluded that chemical fertilizers played an important role in the production of industrial crops in Iran. Table 5.19 also shows that time was not statistically significant in the models.

According to Models 1, 5, 9, 10 and 11, the production process had decreasing returns to scale (DRTS). Based on Models 2, 3, 4, 6, 7, 9 and 12 the returns to scale increased and were equal to 1.091, 1.015, 1.299, 1.020, 1.022, 1.61 and 1.091 respectively.

According to Table 5.19 the use of more chemical fertilizers than animal fertilizers led to a significant decrease in technical efficiency in sugar beet production. The results also show that increased mechanization (defined as use of machinery) led to a significant increase in the technical efficiency of sugar beet production in Iran.

5.3.2.2   Technical Efficiency Measurement

Descriptive statistics for mean technical efficiency according to different models are presented in Table 5.20. The mean technical efficiency of Model 8 (0.932) was the highest, while it was the least for Model 4 (0.248). The average technical efficiency of Models 1, 2, 4 and 5 was about 0.562, 0.665, 0.248 and 0.890 respectively with a maximum value of 1, while the minimum for these models was 0.348, 0.484, 0.118 and 0.526 respectively. Therefore, based on these models some of the provinces were technically efficient. According to Model 4 the gap between the most efficient and inefficient provinces was about 0.882. In Model 8, minimum and maximum technical efficiency scores were 0.79 and 0.98 respectively. Therefore, this model had the least efficiency gap (only 9%). Another result which can be obtained from Table 5.20 is that models in group 3 produced larger efficiency levels as compared to other groups. According to all the models (except for Model 4), mean technical efficiency of sugar beet producing provinces in Iran was higher than 54%. Models in the first group produced similar technical efficiency; mean technical efficiency in this group ranged from 0.546 to 0.566.

Table 5.20 Descriptive statistics for technical efficiency measures by different parametric models

Technical efficiency	Mean	Std. dev.	Minimum	Maximum
Model 1	0.562	0.175	0.348	1
Model 2	0.665	0.142	0.484	1
Model 3	0.546	0.129	0.376	0.869
Model 4	0.248	0.122	0.118	1
Model 5	0.890	0.161	0.526	1
Model 6	0.607	0.161	0.245	0.944
Model 7	0.606	0.161	0.245	0.944
Model 8	0.932	0.042	0.791	0.981
Model 9	0.686	0.085	0.553	0.945
Model 10	0.914	0.109	0.541	0.997
Model 11	0.561	0.175	0.348	0.999
Model 12	0.622	0.111	0.375	0.988

Notes:
Models 1–3: Models with time-invariant inefficiency effects
Models 4–7: Models with time-variant inefficiency effects
Models 8–10: Models separating inefficiency and unobserved individual effects
Models 11–12: Models separating persistent inefficiency from unobserved individual effects

Figure 5.10a and b present kernel density distribution of technical efficiency estimates for Models 1–12 (overall technical efficiency for Models 11 and 12). According to Models 1 and 2, the technical efficiency of a majority of the provinces was about 50%. As an example, consider Models 9 and 10 in which scores for distribution of technical efficiency ranged from 55 to 94 and 54 to 99% respectively. Distribution of technical efficiency in Models 1 to 3 and also in Model 11 had almost the same pattern.

5.3.2.3 Technical Efficiency Measurement by Provinces

Descriptive statistics for technical efficiency measured by provinces are presented in Table 5.21. Estimated technical efficiency according to Models 1 and 2 indicates that West Azerbaijan was technically fully efficient with a value of 100%. According to all the models (except for Models 9 and 10), Kerman was the least efficient province in sugar beet production. Technical efficiency scores for this province based on Models 1 to 8 are 0.348, 0.484, 0.376, 0.153, 0.830, 0.421 and 0.870 respectively.

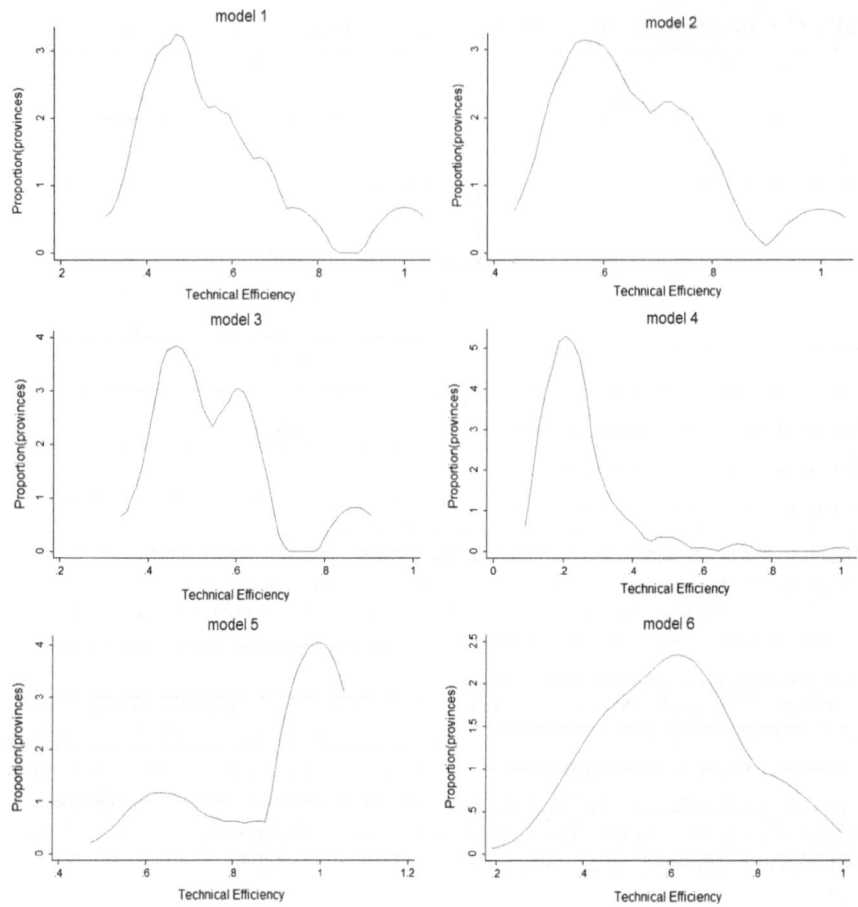

Fig. 5.10 a Technical efficiency distributions of sugar beet producing provinces for Models 1–6,
b Technical efficiency distributions of sugar beet producing provinces for Models 7–12

Estimated efficiency scores for Kermanshah reveal that this province was one of
the efficient provinces in sugar beet production; its technical efficiency scores were
higher than 60% for all the models. According to this table, Isfahan was one of the
least efficient provinces. Based on Model 5, Qazvin's technical efficiency was 89%.
These results are consistent with Yazdani and Rahimi (2012) who concluded that
sugar beet production in Qazvin had a technical efficiency of 89.6%.

Table 5.22 gives the ranks of different provinces by the level of their efficiency.
According to most of the models, West Azerbaijan was the most technically efficient
province in sugar beet production in Iran. According to statistics, this province had
the highest yield and largest harvested area of sugar beet in the country. Models 1, 2,
3, 6, 7, 11 and 12 calculated almost the same ranking for the provinces and concluded
that West Azerbaijan, Lorestan and Kermanshah were the most efficient. According

Table 5.21 Descriptive statistics for technical efficiency measures by provinces

Province	Model 1	Model 2	Model 3	Model 4	Model 5	Model 6	Model 7	Model 8	Model 9	Model 10	Model 11	Model 12
Markazi	0.473	0.632	0.499	0.232	0.865	0.549	0.549	0.925	0.704	0.983	0.472	0.665
West Azerbaijan	1.000	1.000	0.869	0.382	0.950	0.907	0.907	0.968	670	0.939	0.999	0.899
Kermanshah	0.623	0.755	0.611	0.329	0.979	0.753	0.753	0.931	0.683	0.974	0.623	0.756
Fars	0.513	0.553	0.483	0.202	0.868	0.549	0.549	0.931	0.667	0.980	0.512	0.550
Kerman	0.384	0.484	0.376	0.153	0.830	0.421	0.421	0.870	0.692	0.938	0.348	0.320
Khorasan	0.740	0.675	0.610	0.231	0.876	0.858	0.585	0.931	0.661	0.890	0.739	0.655
Isfahan	0.421	0.533	0.433	0.185	0.858	0.504	0.504	0.899	0.689	0.776	0.421	0.503
Chaharmahal and Bakhtiari	0.443	0.582	4640	0.245	0.842	0.515	0.515	0.968	0.689	0.950	0.443	0.542
Lorestan	0.623	0.798	0.632	0.298	0.944	0.722	0.722	0.931	0.689	0.949	0.622	0.680
Semnan	0.442	0.578	0.462	0.213	0.878	0.532	0.532	0.941	0.693	0.700	0.442	0.570
Qazvin	0.551	0.726	0.568	0.258	0.896	0.644	0.644	0.963	0.693	0.978	0.551	0.703

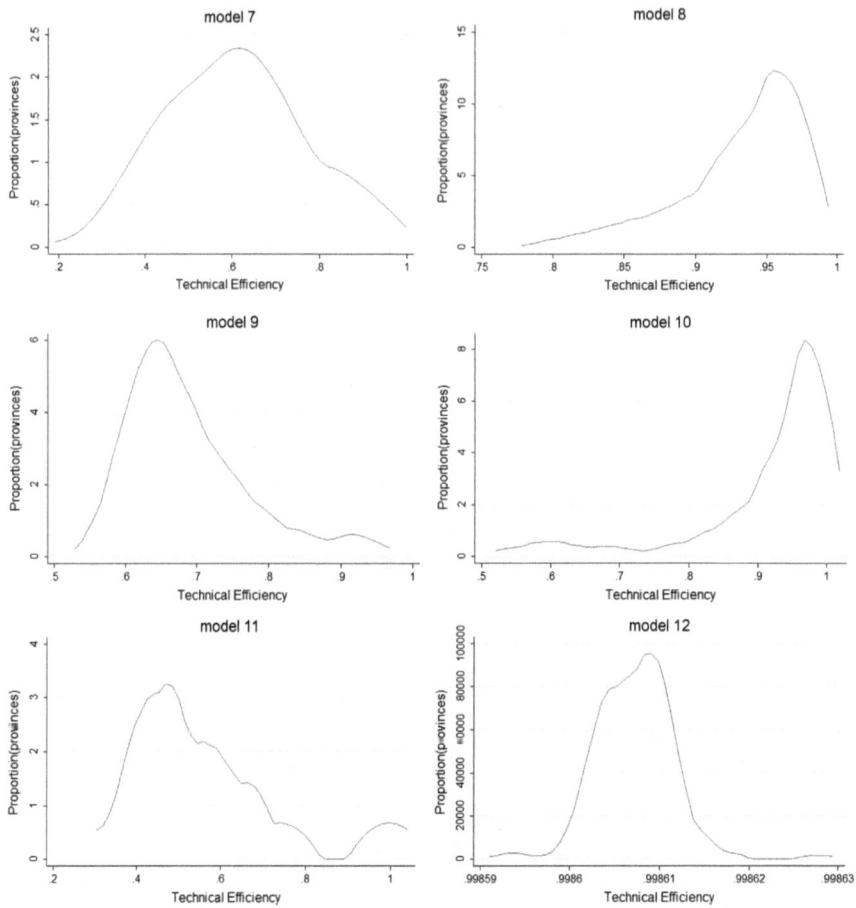

Fig. 5.10   (continued)

to these models, Kerman and Isfahan were the least efficient. Markazi, which had the minimum harvested area and production of sugar beet in the country, ranked sixth and seventh respectively in technical efficiency of production.

5.3.2.4   Technical Efficiency Measurement by Years

Mean provincial technical efficiency for the 12 models during 2000–2012 is presented in Table 5.23. As expected technical efficiency for Models 1–3 was constant during study period. According to Model 4, technical efficiency increased slightly during 2000–2012. Models 8 and 10 present less fluctuations as compared to other models. According to all the models (except for Models 4 and 9), mean technical efficiency scores decreased during the study period.

Table 5.22 Ranking sugar beet producing provinces by different parametric models

Rank	Model 1	Model 2	Model 3	Model 4	Model 5	Model 6	Model 7	Model 8	Model 9	Model 10	Model 11	Model 12
1	West Azerbaijan	West Azerbaijan	West Azerbaijan	West Azerbaijan	Kermanshah	West Azerbaijan	West Azerbaijan	West Azerbaijan	Markazi	Markazi	West Azerbaijan	West Azerbaijan
2	Khorasan	Lorestan	Lorestan	Kermanshah	West Azerbaijan	Kermanshah	Kermanshah	Chaharmahal and Bakhtiari	Lorestan	Fars	Khorasan	Lorestan
3	Kermanshah	Kermanshah	Kermanshah	Lorestan	Lorestan	Lorestan	Lorestan	Qazvin	Chaharmahal and Bakhtiari	Qazvin	Kermanshah	Kermanshah
4	Lorestan	Qazvin	Khorasan	Qazvin	Qazvin	Qazvin	Qazvin	Semnan	Semnan	Kermanshah	Lorestan	Qazvin
5	Qazvin	Khorasan	Qazvin	Chaharmahal and Bakhtiari	Semnan	Khorasan	Khorasan	Kermanshah	Qazvin	Chaharmahal and Bakhtiari	Qazvin	Khorasan
6	Fars	Markazi	Markazi	Markazi	Khorasan	Fars	Fars	Fars	Kerman	Lorestan	Fars	Markazi
7	Markazi	Chaharmahal and Bakhtiari	Fars	Khorasan	Fars	Markazi	Markazi	Khorasan	Isfahan	West Azerbaijan	Markazi	Chaharmahal and Bakhtiari
8	Chaharmahal and Bakhtiari	Semnan	Chaharmahal and Bakhtiari	Semnan	Markazi	Semnan	Semnan	Lorestan	Kermanshah	Kerman	Chaharmahal and Bakhtiari	Semnan
9	Semnan	Fars	Semnan	Fars	Isfahan	Chaharmahal and Bakhtiari	Chaharmahal and Bakhtiari	Markazi	West Azerbaijan	Khorasan	Semnan	Fars
10	Isfahan	Isfahan	Isfahan	Isfahan	Chaharmahal and Bakhtiari	Isfahan	Isfahan	Isfahan	Fars	Isfahan	Isfahan	Isfahan
11	Kerman	Kerman	Kerman	Kerman	Kerman	Kerman	Kerman	Kerman	Khorasan	Semnan	Kerman	Kerman

Table 5.23 Descriptive statistics for technical efficiency measures by years

Year	Model 1	Model 2	Model 3	Model 4	Model 5	Model 6	Model 7	Model 8	Model 9	Model 10	Model 11	Model 12
2000	0.562	0.660	0.546	0.205	1.000	0.738	0.738	0.924	0.686	0.964	0.570	0.630
2001	0.562	0.660	0.546	0.196	1.000	0.719	0.719	0.906	0.667	0.935	0.554	0.614
2002	0.562	0.660	0.546	0.191	1.000	0.700	0.700	0.922	0.637	0.912	0.563	0.625
2003	0.562	0.660	0.546	0.190	1.000	0.680	0.680	0.925	0.636	0.920	0.555	0.627
2004	0.562	0.660	0.546	0.192	1.000	0.658	0.658	0.932	0.664	0.905	0.562	0.630
2005	0.562	0.660	0.546	0.199	1.000	0.636	0.636	0.926	0.648	0.889	0.576	0.642
2006	0.562	0.660	0.546	0.209	1.000	0.613	0.613	0.940	0.621	0.907	0.564	0.616
2007	0.562	0.660	0.546	0.223	1.000	0.589	0.589	0.948	0.648	0.934	0.567	0.634
2008	0.562	0.660	0.546	0.243	0.799	0.564	0.564	0.939	0.803	0.908	0.583	0.647
2009	0.562	0.660	0.546	0.271	0.692	0.539	0.539	0.923	0.788	0.954	0.596	0.665
2010	0.562	0.660	0.546	0.309	0.693	0.513	0.513	0.956	0.709	0.887	0.538	0.570
2011	0.562	0.660	0.546	0.361	0.693	0.487	0.487	0.946	0.693	0.885	0.566	0.615
2012	0.562	0.660	0.546	0.434	0.692	0.460	0.460	0.937	0.723	0.893	0.551	0.572

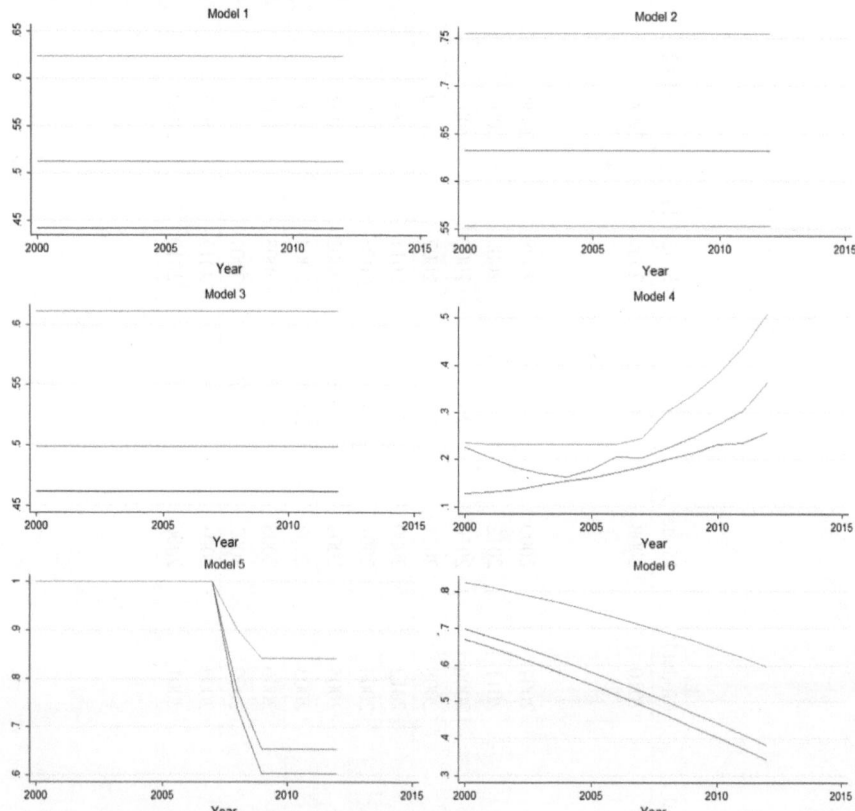

Fig. 5.11 **a** The mean, first and third quartile values of technical efficiency of different sugar beet producing provinces for Model 1–6, **b** The mean, first and third quartile values of technical efficiency of different sugar beet producing provinces for Model 7–12

Table 5.24 gives the rankings of different years in the study period. According to Models 5–7, 2000 was the most efficient year and 2012 the least efficient year during the study period.

Figure 5.11 shows the lower quartile, median and upper quartile efficiencies over time across the 12 models. According to this, Models 1–3 had the widest efficiency spread and Models 8 and 9 had the narrowest spread. Mean technical efficiency of Models 1 and 3 were almost the same (Table 5.24).

Pairwise rank-order correlations for technical efficiency scores of Models 1 through 12 are reported in Table 5.25. Model 1 had a positive and significant correlation with the other models (except for Models 5 and 9). The rank-order correlation of this model with Models 2, 3 and 11 was high. The correlation between Models 6 and 7 ($\tau = 1.000$), and also between Models 2 and 3 ($\tau = 0.816$) was high. Therefore, these models seem to be consistent in generating similar results. We found the

Table 5.24 Ranking of different years

Rank	Model 4	Model 5	Model 6	Model 7	Model 8	Model 9	Model 10	Model 11	Model 12
1	2012	2000, 2001, 2002, 2003, 2004, 2005, 2006	2000	2000	2010	2008	2000	2009	2009
2	2011	2008	2001	2001	2007	2009	2009	2008	2008
3	2010	2009	2002	2002	2011	2012	2001	2005	2005
4	2009	2010	2003	2003	2006	2010	2007	2000	2007
5	2008	2011	2004	2004	2008	2011	2003	2007	2000
6	2007	2012	2005	2005	2012	2000	2002	2011	2004
7	2006	–	2006	2006	2004	2001	2008	2006	2003
8	2000	–	2007	2007	2005	2004	2006	2002	2002
9	2005	–	2008	2008	2003	2007	2004	2004	2006
10	2001	–	2009	2009	2000	2005	2012	2003	2011
11	2004	–	2010	2010	2009	2002	2005	2001	2001
12	2002	–	2011	2011	2002	2003	2011	2012	2012
13	2003	–	2012	2012	2001	2006	2010	2010	2010

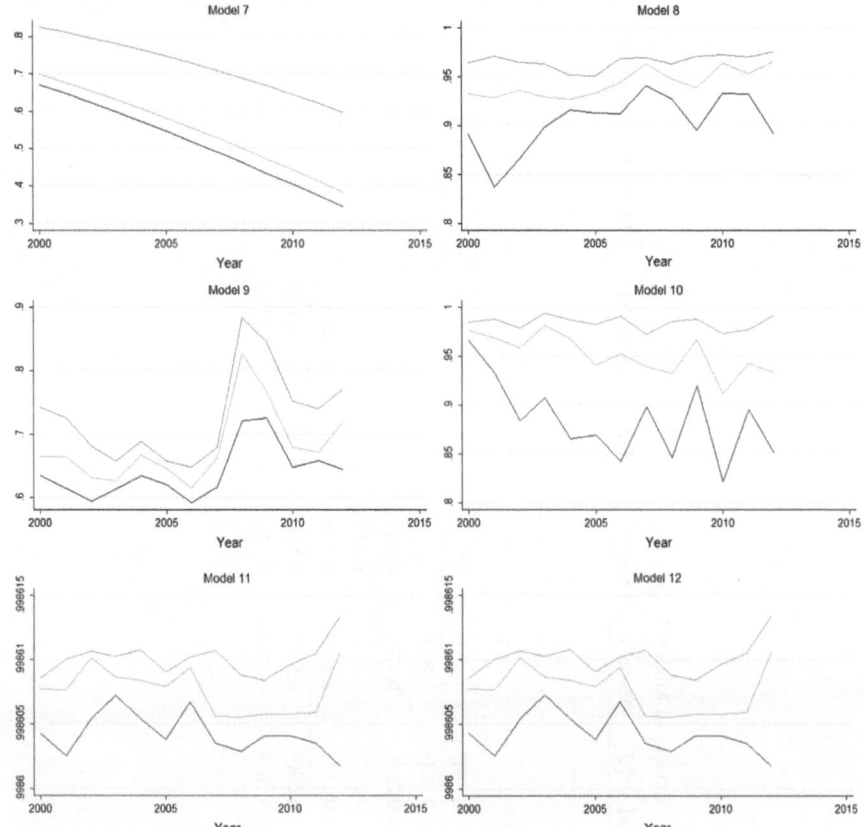

Fig. 5.11 (continued)

same results for the cotton crop (see Table 5.9). In contrast, the correlation between Models 4 and 5, Models 5 and 8, Models 9 and 5 and Models 6 and 7 were negative.

Table 5.26 shows Kendall's rank-order correlation for persistent and residual technical efficiency of Models 11 and 12. According to this table, assessments of residual technical efficiency for these two models were to a large extent positively correlated with a rank-order correlation of 0.755. The results also indicate that persistent and residual technical efficiency of Model 11 was negatively correlated (this result confirms our findings in cotton production). This table shows that the correlation between persistent technical efficiency of Models 11 and 12 was high.

In Fig. 5.12 scatter plot matrices for Models 1–12 graphically illustrate the differences between the models in the ranking of the provinces. These results prove the findings given in Table 5.25.

Table 5.25 Kendall's rank-order correlation between the technical efficiency of Models 1–12

	Model 1	Model 2	Model 3	Model 4	Model 5	Model 6	Model 7	Model 8	Model 9	Model 10	Model 11
Model 2	0.683[a]										
Model 3	0.782[a]	0.816[a]									
Model 4	0.324[a]	0.395[a]	0.363[a]								
Model 5	0.079	0.084[a]	0.081[a]	−0.233[a]							
Model 6	0.505[a]	0.549[a]	0.547[a]	0.056	0.381[a]						
Model 7	0.505[a]	0.549[a]	0.547[a]	0.056	0.381[a]	1.000[a]					
Model 8	0.238[a]	0.311[a]	0.247[a]	0.414[a]	−0.068	0.133	0.133				
Model 9	−0.010	0.053	0.024	0.244[a]	−0.293[a]	−0.117[a]	−0.117[a]	0.069			
Model 10	0.094[a]	0.098[a]	0.119[a]	0.020	0.101[a]	0.146[a]	0.146[a]	0.063	−0.037		
Model 11	0.916[a]	0.683[a]	0.782[a]	0.326[a]	0.090[a]	0.510[a]	0.510[a]	0.267[a]	−0.011	0.099[a]	
Model 12	0.160[a]	0.178[a]	0.174[a]	0.182[a]	0.101[a]	0.176[a]	0.176[a]	0.287[a]	0.106[a]	0.051	0.231[a]

Notes: Significant at less than 1% ([a])

Table 5.26 Kendall's rank-order correlation between the technical efficiency of Models 11 and 12 for persistent technical efficiency (PTE) and residual technical efficiency (RTE) estimates

	Model 11 PTE	Model 11 RTE	Model 12 PTE
Model 11 RTE	−0.002		
Model 12 PTE	0.682[a]	0.002	
Model 12 RTE	0.155[a]	0.755[a]	0.172[a]

Notes: Significant at less than 1% ([a])

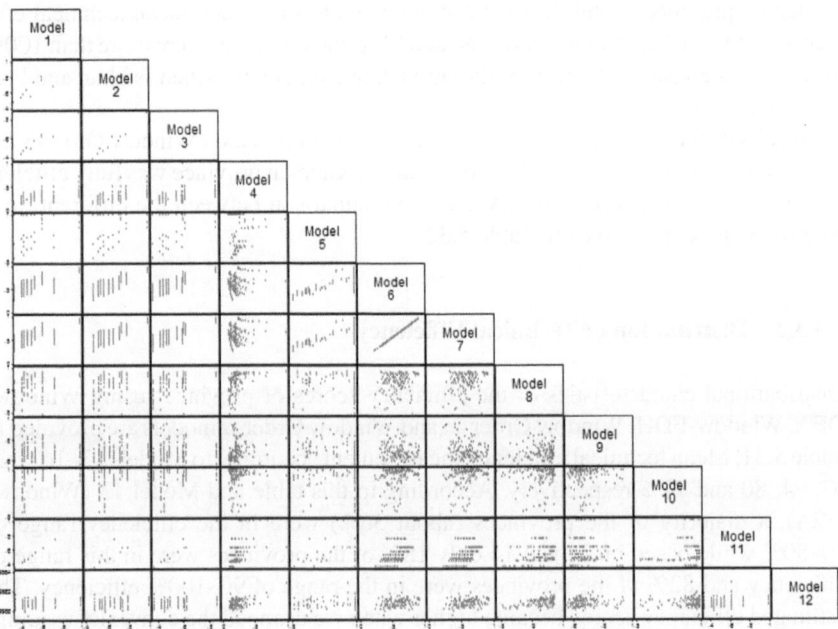

Fig. 5.12 Scatter plot matrices of pairwise technical efficiency estimates for Models 1–12

5.3.3 Non-parametric Models

5.3.3.1 Provincial Technical Efficiency Measurement

Tables 5.27, 5.28, 5.29 and 5.30 give the mean efficiency scores obtained using Window-DEA, Window-FDH, Window Order-m and Window Order-α models for different sugar beet producing provinces during 2000–2012. First, we explain the efficiency scores obtained from the Window-DEA model in Table 5.27. According to this model West Azerbaijan, Khorasaan and Lorestan were the most efficient sugar beet producing provinces in Iran while Chaharmahal and Bakhtiari and Fars were the least efficient provinces. According to this table, Khorasan province's technical efficiency decreased during most years of the study period. Results of Models 14–16 in Tables 5.28 and 5.30 show the same technical efficiency distribution for Khorasan.

Lorestan had an increasing trend in efficiency during 2000–2005, after which it experienced huge fluctuations from 2005 to 2012.

With respect to the Window-FDH model, Table 5.28 shows that West Azerbaijan, Lorestan and Khorasan were the most efficient provinces which confirms the results of the Window-DEA model. These provinces were fully efficient in most of the years. According to Table 5.28, Fars and Isfahan provinces were the least efficient in sugar beet production in Iran.

According to Table 5.13 which presents mean technical efficiency of sugar beet producing provinces obtained by the Window Order-m model, mean technical efficiencies of West Azerbaijan, Khorasan and Lorestan provinces were more than 100% in some of the years. As a full frontier model, this model classified Isfahan and Fars as the least efficient provinces.

Results of mean technical efficiency measurement based on Window Order-α are presented in Table 5.30. According to this table, Khorsan province was fully efficient during 2000–2007 and in 2012. A detailed comparison between the different non-parametric models is given in Table 5.32.

5.3.3.2   Distribution of Technical Efficiency

Distributional characteristics of the efficiency scores of provinces using Window-DEA, Window-FDH, Window Order-m and Window Order-α models are provided in Table 5.31. Mean technical efficiency measurement according to Models 13–16 were 67, 94, 80 and 77% respectively. According to this table and Model 13 (Window-DEA), a majority of the provinces (about 36%) were in the efficiency range of 70–80% while based on Model 14 only 18% of the provinces were in this range of efficiency and 82% of the provinces were in the range of 90–100% efficiency. The estimated efficiency scores became higher when we dropped the convexity assumption in non-parametric models. Based on Model 15, minimum and maximum technical efficiency scores were 61 and 104% respectively. Hence, the gap between the most efficient and least efficient provinces was about 43%. Model 16 reports 56 and 101% minimum and maximum technical efficiency. It shows the highest efficiency gap (about 45%) among sugar beet producing provinces which proves regional differences in terms of allocating resources and inputs for sugar beet production in Iran. According to Models 15 and 16, about 9.09% of the provinces were in the range of equal to or larger than 1.00 efficiency and was considered as super-efficient provinces. For comparison with other approaches, if necessary, the number can be normalized to a maximum 1.0.

5.3.3.3   Ranking of Different Sugar Beet Producing Provinces

Table 5.32 presents sugar beet producing provinces' ranks determined by different non-parametric models. According to this table, full frontier models produced the same results and found West Azerbaijan to be the most efficient province. Accord-

Table 5.27 Mean technical efficiency of different windows during 2000–2012 (Window-DEA model)

Province	2000	2001	2002	2003	2004	2005	2006	2007	2008	2009	2010	2011	2012
Markazi	0.65	0.62	0.54	0.63	0.52	0.71	0.79	0.60	0.93	0.76	0.62	0.60	0.58
West Azerbaijan	1.00	0.90	0.98	0.99	1.00	0.65	0.99	0.94	0.75	0.70	0.72	0.63	0.55
Kermanshah	0.67	0.31	0.78	0.62	0.61	0.97	0.83	1.00	0.74	0.82	0.59	1.00	0.52
Fars	0.42	0.45	0.48	0.58	0.54	0.55	0.60	0.43	0.71	0.48	0.45	0.62	1.00
Kerman	0.56	0.42	0.48	1.00	1.00	0.51	0.62	0.59	1.00	0.41	0.94	0.37	0.52
Khorasan	0.86	0.84	0.92	0.89	0.78	0.80	0.86	0.71	0.65	0.87	0.53	0.64	0.38
Isfahan	0.42	0.42	0.46	0.63	0.86	0.61	0.58	0.83	0.52	0.61	0.51	0.54	0.31
Chaharmahal and Bakhtiari	0.45	0.54	0.66	0.58	0.58	0.56	0.54	0.64	0.44	0.53	0.67	0.34	0.61
Lorestan	0.48	0.56	0.92	0.82	0.86	1.00	0.86	0.53	0.75	0.85	0.70	0.97	0.34
Semnan	0.79	0.57	0.63	0.60	0.61	0.47	0.61	0.56	0.46	0.62	0.90	0.42	0.60
Qazvin	0.66	1.00	0.96	0.78	0.65	0.84	0.99	0.78	0.80	0.55	0.55	0.50	0.50
Mean	0.63	0.60	0.71	0.74	0.73	0.70	0.75	0.69	0.70	0.65	0.65	0.60	0.54

Table 5.28 Mean technical efficiency of different windows during 2000–2012 (Window-FDH model)

	2000	2001	2002	2003	2004	2005	2006	2007	2008	2009	2010	2011	2012
Markazi	1.00	0.99	0.79	0.91	0.85	1.00	1.00	1.00	1.00	0.94	0.94	1.00	0.67
West Azerbaijan	1.00	0.98	1.00	1.00	1.00	0.94	1.00	1.00	1.00	0.98	0.98	1.00	1.00
Kermanshah	1.00	0.43	1.00	1.00	1.00	1.00	1.00	1.00	1.00	0.99	0.80	0.87	1.00
Fars	0.68	0.57	0.75	0.95	0.90	0.91	1.00	0.69	1.00	0.75	0.94	1.00	0.96
Kerman	0.96	0.79	0.94	1.00	1.00	0.73	0.98	1.00	1.00	0.88	1.00	0.95	0.80
Khorasan	1.00	1.00	1.00	1.00	1.00	1.00	1.00	1.00	1.00	0.92	0.90	0.91	1.00
Isfahan	1.00	0.84	0.75	0.87	1.00	0.72	0.93	1.00	0.83	1.00	0.85	1.00	0.62
Chaharmahal and Bakhtiari	1.00	0.89	1.00	1.00	0.99	1.00	1.00	1.00	1.00	1.00	1.00	0.79	1.00
Lorestan	1.00	0.92	1.00	1.00	1.00	1.00	1.00	1.00	1.00	1.00	0.93	0.95	1.00
Semnan	1.00	1.00	0.88	0.97	1.00	0.83	1.00	0.84	0.92	0.96	0.93	0.75	1.00
Qazvin	0.68	1.00	1.00	1.00	1.00	1.00	1.00	1.00	0.91	0.98	0.77	0.89	1.00
Mean	0.94	0.86	0.92	0.97	0.98	0.92	0.99	0.96	0.97	0.95	0.91	0.92	0.91

Table 5.29 Mean technical efficiency of different windows during 2000–2012 (Window Order-m model)

	2000	2001	2002	2003	2004	2005	2006	2007	2008	2009	2010	2011	2012
Markazi	1.00	0.90	0.55	0.66	0.67	0.71	0.84	1.08	1.10	1.02	0.92	1.14	0.76
West Azerbaijan	0.86	0.83	0.97	0.76	1.00	0.77	1.00	0.88	1.00	0.92	0.95	1.00	1.00
Kermanshah	0.34	0.14	0.25	0.32	1.00	0.70	1.00	0.48	0.77	1.01	0.47	0.81	1.77
Fars	0.58	0.45	0.39	0.71	0.66	0.51	0.57	0.61	1.09	0.46	0.82	1.09	0.96
Kerman	0.56	0.59	0.52	1.33	1.30	0.64	0.68	0.66	1.59	0.77	2.25	1.26	1.39
Khorasan	1.00	1.00	1.00	1.00	1.00	1.00	1.00	1.00	0.63	0.67	0.79	0.83	1.02
Isfahan	0.43	0.29	0.19	0.30	0.58	0.48	0.28	0.69	0.94	1.17	0.70	1.05	0.78
Chaharmahal and Bakhtiari	0.59	0.72	0.55	0.51	0.53	0.49	0.42	0.59	1.50	2.47	1.11	0.77	1.21
Lorestan	0.68	0.68	0.85	0.85	0.60	0.73	0.43	0.33	1.08	1.32	0.86	0.96	1.27
Semnan	0.77	0.49	0.46	0.50	0.59	0.40	0.45	0.58	0.67	0.91	0.93	0.60	1.28
Qazvin	0.33	0.50	0.40	0.41	0.60	0.84	1.00	0.72	1.06	0.98	0.68	0.87	1.10
Mean	0.65	0.60	0.56	0.67	0.77	0.66	0.70	0.69	1.04	1.06	0.95	0.94	1.14

Table 5.30 Mean technical efficiency of different windows during 2000–2012 (Window Order-α model)

	2000	2001	2002	2003	2004	2005	2006	2007	2008	2009	2010	2011	2012
Markazi	1.00	0.94	0.57	0.53	0.72	0.71	0.95	1.07	1.09	1.00	0.94	1.12	0.83
West Azerbaijan	0.86	0.83	0.97	0.76	1.00	0.77	1.00	0.88	1.00	0.91	0.95	1.00	1.00
Kermanshah	0.27	0.12	0.23	0.23	1.00	0.62	1.00	0.37	0.66	0.97	0.40	0.75	1.00
Fars	0.58	0.45	0.39	0.71	0.65	0.51	0.57	0.61	1.06	0.42	0.81	1.00	0.96
Kerman	0.47	0.57	0.61	1.56	1.43	0.67	0.63	0.72	2.10	0.79	1.59	1.00	1.00
Khorasan	1.00	1.00	1.00	1.00	1.00	1.00	1.00	1.00	0.63	0.66	0.79	0.82	1.00
Isfahan	0.36	0.25	0.17	0.25	0.54	0.44	0.23	0.56	1.00	1.25	0.55	1.00	0.62
Chaharmahal and Bakhtiari	0.64	0.73	0.43	0.42	0.44	0.38	0.35	0.49	1.85	2.60	1.11	0.80	1.21
Lorestan	0.64	0.72	0.72	0.76	0.55	0.63	0.38	0.29	1.04	1.57	0.78	0.91	1.00
Semnan	0.62	0.39	0.38	0.45	0.56	0.35	0.40	0.52	0.70	0.94	0.79	0.57	1.31
Qazvin	0.30	0.45	0.35	0.36	0.52	0.87	0.97	0.66	1.00	1.00	0.63	0.79	1.00
Mean	0.61	0.59	0.53	0.64	0.76	0.63	0.68	0.65	1.10	1.10	0.85	0.89	0.99

Table 5.31 Descriptive statistics for the average efficiency scores by different non-parametric models

Intervals	Model 13	Model 14	Model 15	Model 16
=> 1.00	–	–	9.09	9.09
0.90–1.00	–	81.82	18.18	18.18
0.80–0.90	0.09	18.18	27.27	18.18
0.70–0.80	36.36	–	9.09	9.09
0.60–0.70	27.27	–	36.36	27.27
0.50–0.60	27.27	–	–	18.18
0.40–0.50	–	–	–	–
0.30–0.40	–	–	–	–
0.20–0.30	–	–	–	–
0.10–0.20	–	–	–	–
0.00–0.10	–	–	–	–
Mean	0.67	0.94	0.80	0.77
Skewness	0.10	0.60	1.90	0.015
Kurtosis	1.76	2.56	0.16	1.54

ing to these models, Fars was the least efficient province. The last two columns of Table 5.32 give the rankings of provinces based on partial frontier models. According to these models, Kerman and West Azerbaijan were the most efficient and Isfahan was the least efficient sugar beet producing province. Based on these two models, Khorasan, the second largest sugar beet producer in the country was also the second highest in terms of technical efficiency (Table 5.33).

5.3.3.4   Kendall's Rank-Order Correlation Between Non-parametric Models

Pairwise rank-order correlations of different models are reported in Table 5.31. According to this table, the correlation between full frontier models was relatively high at 0.69, indicating that these two models ranked different provinces in the same order. Meanwhile, as expected, the correlation among Model 13 and Models 15 and 16 are insignificant. An insignificant correlation was found between Models 14 and 15, and also among Models 14 and 16. The rank-order correlation between two partial frontier models (Order-m and Order-α) was significant and very high (0.96).

Table 5.32 Scores and ranking values obtained by non-parametric models

Provinces	Model 13 window DEA scores	Model 14 window FDH scores	Model 15 window order-m scores	Model 16 window order-α scores	Model 13 window-DEA ranking	Model 14 window FDH ranking	Model 15 window order-m ranking	Model 16 window order-α ranking
Markazi	0.66	0.93	0.87	0.88	6	7	4	3
West Azerbaijan	0.83	0.99	0.92	0.92	1	1	2	2
Kermanshah	0.73	0.93	0.70	0.59	5	6	7	8
Fars	0.56	0.85	0.69	0.67	10	11	8	6
Kerman	0.65	0.92	1.04	1.01	7	9	1	1
Khorasan	0.75	0.98	0.92	0.92	2	3	2	2
Isfahan	0.56	0.88	0.61	0.56	9	10	10	9
Chaharmahal and Bakhtiari	0.55	0.97	0.88	0.88	11	4	3	3
Lorestan	0.74	0.98	0.82	0.77	3	2	5	4
Semnan	0.60	0.93	0.66	0.61	8	8	9	7
Qazvin	0.74	0.94	0.73	0.68	4	5	6	5
Mean	0.67	0.94	0.80	0.77	–	–	–	–

Table 5.33 Kendall's rank-order correlation between technical efficiency of non-parametric models

	Model 13	Model 14	Model 15	Model 16
Model 13	1			
Model 14	0.69[a]	1		
Model 15	0.44	0.53[a]	1	
Model 16	0.42	0.47	0.96[a]	1

Notes: Significant at less than 1% ([a])

Chapter 6
Analysis of Results and Policy Recommendations

Based on the analysis of the empirical results in earlier chapters, this chapter offers some recommendations for both policymakers and producers of industrial crops in Iran.

Since Khorasan province emerged as the most efficient province in Iran's cotton production, it is recommended that the government treat this province as a model when distributing physical resources for optimal production and efficiency in the rest of the country.

Large variations in the level of technical efficiency in the provinces indicate large disparities among the sample provinces in terms of technical efficiency. This might be due to the differences in comparative advantages of some provinces in industrial crop production. According to the results, West Azerbaijan and Khorasan provinces have comparative advantage in sugar beet production. Therefore, specialization in the use of resources should be promoted to increase production of this industrial crop in the country.

Since the level of technical efficiency in the sample provinces declined over time, designing appropriate policies for improving efficiency at the farm level is of utmost importance. In doing so, great emphasis should be placed on strengthening the capacity of cotton and sugar beet farmers through organizing farmer training workshops geared towards improvements in managerial and resource use efficiencies. This should be done in a collaborative manner involving the government, district assemblies and non-governmental organizations.

Large differences between the sample units show provincial disparities, which in turn indicate that location has a significant impact on the efficiency of these industrial crops in Iran. One reason for this could be the optimality of conditions in the production environment for industrial crop production that exist in different parts of the country. This implies further research which considers geographical conditions in measuring productivity and technical efficiency in different production zones.

Inefficiency in the farming of industrial crops can be reduced significantly using efficient machinery at different times of the production process to reduce labor-use

© Springer Nature Singapore Pte Ltd. 2018   115
M. Rashidghalam, *Measurement and Analysis of Performance of Industrial Crop Production:*
The Case of Iran's Cotton and Sugar Beet Production, Perspectives on Development in the Middle
East and North Africa (MENA) Region, https://doi.org/10.1007/978-981-13-0092-9_6

intensity in the production of these crops. Therefore, it is recommended that more attention be paid to this aspect to increase efficiency and to contribute to increased factor productivity and output growth in the country.

Since Model 12 is a generalized version of other parametric efficiency models, it is recommended that future researchers use this model for measuring crops' technical efficiency especially if the objective of their research is studying and analyzing the differences between persistent inefficiency and time-varying inefficiency. Separating persistent and transitory components of production inefficiency from individual effects will allow the state to distinguish between long- and short-term perspectives on general modernization of agriculture and targeted efficiency and productivity enhancements. This information can be useful in the allocation of resources and in designing public investment policies and programs.

A low level of technical efficiency indicates the presence of significant opportunities for improvements in industrial crop production in Iran which could improve productivity of all inputs in multi-input production processes. A challenge here will be building a suitable marketing system for increased production, especially for sugar beet. A combination of production, processing and marketing in policy will be conductive to the sector's progress and efficiency.

This research had limited information on technical efficiency determinants of cotton and sugar beet producers in the country (only data on the percentage of agricultural machinery utilization was available). Therefore, it is recommended that the Ministry of Agriculture, Jihad should provide researchers with panel data containing information covering different regions of the country. This information should include extension services and training of farmers in each province, farm demographics and information on human capital, information technology access and use, capital investments and level of farm production technology in each province.

Given that the non-parametric partial frontier models have assumptions which are closer to reality as compared to full frontier models, it is recommended that future studies use partial frontier models rather than full frontier ones for measuring technical efficiency and for ranking decision making units (DMUs).

Regarding partial frontier models, this book used the values of $m = 21$ and $\alpha = 95$ for technical efficiency measurement in Order-m and Order-α models, respectively. It is desirable that future works in this area cover different values for m (for example, $m = 75, 150, 300$ and $1,500$) and α (for example, $\alpha = 98.0, 98.5, 99.0$ and 99.5) to study how the results of technical efficiency scores and ranking of DMUs vary in response to different chosen values. Our findings reflect that the results are sensitive to modeling and parameter size selections which need to be accounted for in the estimation of efficiency.

Chapter 7
Summary and Conclusion

7.1 Overview

This book used both parametric and non-parametric up-to-date and commonly used models to evaluate technical efficiency of Iranian provinces in industrial crop production (cotton and sugar beet). It also discussed the determinants of technical efficiency to be used in policymaking to support industries. This book's secondary goal was ranking different provinces and determining efficient provinces in cotton and sugar beet production in Iran so that these can serve as models while allocating public support to the provinces for enhancing fairness and efficiency in society.

Chapter 1 discussed the introduction and main goals, assumptions and questions of the study. Chapter 2 introduced the status of agricultural industrial crops in Iran with focus on cotton and sugar beet production. Chapter 3 discussed the theoretical framework of efficiency measurement and also gave a literature review. A complete description of 12 parametric and four non-parametric advanced models frequently used for the study were introduced in Chap. 4. The study of parametric models was classified into four groups in terms of the assumptions made on the temporal behavior of inefficiency. A common issue among all the parametric models is that inefficiency was individual producer-specific. This is consistent with the notion of measuring the efficiency of decision making units. Group 1, which assumed the inefficiency effects to be time-invariant and individual-specific includes Models 1–3. Models 4–7 in Group 2 allowed inefficiency to be individual-specific but also time-varying. In group 3, Models 8–10 separated inefficiency effects from unobserved individual effects. Finally, in group 4, Model 11 and 12 separated persistent inefficiency and time-varying inefficiency from unobservable individual effects.

The second part of Chap. 4 detailed the four non-parametric models which were divided into partial and full frontier models. A Window analysis was done to consider the effect of time in these models. Although the data envelopment analysis (DEA) and the free disposable hull (FDH) methods have some advantages over parametric

M. Rashidghalam, *Measurement and Analysis of Performance of Industrial Crop Production: The Case of Iran's Cotton and Sugar Beet Production*, Perspectives on Development in the Middle East and North Africa (MENA) Region, https://doi.org/10.1007/978-981-13-0092-9_7

methods they also have some limitations and have been criticized by econometricians for lacking a well-defined data generating process, being deterministic and being highly sensitive to measurement errors and extreme data observations. This book addressed some of these objections by introducing partial frontier approaches Order-m and Order-α. These two techniques generalize the FDH model by allowing super-efficient DMUs to be located beyond the estimated production possibility frontier. Therefore, two partial frontier models were considered and their results compared with full frontier models using a Window analysis.

A Window analysis is a variation of the traditional non-parametric approach and handles cross-sectional and time-varying data to measure dynamic effects. This technique is based on the principle of moving averages and establishes efficiency measures by treating each DMU in different periods as a separate unit.

Finally, Chap. 5 employed the models for cotton and sugar beet producing provinces in Iran. This chapter included 26 models measuring technical efficiency. It also provided a comparison of the different models.

7.2   Summary of Empirical Results

7.2.1   Cotton Production

To evaluate technical efficiency of cotton producing provinces in Iran, information on inputs and outputs from Markazi, Mazandaran, East Azerbaijan, Fars, Kerman, Khorasan, Isfahan, Semnan, Yazd, Tehran, Golestan, Ardabil and Qom provinces was collected for the period 2000–2012. Khorasan and Golestan provinces are the most important provinces in cotton crop production in Iran.

The results showed that pesticides, chemical fertilizers, animal fertilizers and labor were important inputs in the production of this crop. The production process had a decreasing returns to scale (DRTS). A study of the determinants of technical efficiency showed that more chemical fertilizers relative to animal fertilizers led to a significant increase in technical efficiency in cotton production. Use of machinery also led to a significant increase in technical efficiency in cotton production in Iran.

A comparison of the nested and non-nested parametric models showed that Model 4 (Cornwell et al. 1990) was the most preferable stochastic frontier panel data model. According to this model, average technical efficiency of cotton producing provinces was about 46% with a maximum value of 100% and a minimum value of only 9.3%. Highest and lowest average technical efficiency was measured for model groups 1 and 3 respectively. According to most of the parametric models Khorasan, Ardabil and East Azerbaijan were the most technically efficient provinces and ranked first to third among the provinces. Based on these models Yazd, Qom and Mazandaran provinces were the least technically efficient cotton producing provinces in Iran. Trends of technical efficiency in production during 2000–2012 indicate that according to all the time-variant models (except Model 9), technical efficiency decreased during the study period. Pairwise rank order correlations indicate that Models 1, 2, 3 and 11 and Models 5, 6 and 7 generated similar results. They ranked the provinces in the same order but not necessarily at the same efficiency level.

The results of the non-parametric models indicate that mean technical efficiency based on Window-DEA, Window-FDH, Window Order-m and Window Order-α were about 68, 72, 91 and 111%. Window Order-α generated super-efficiency observations. A normalization will enable comparability with the parametric frontier models' results. According to full frontier models (DEA and FDH), Khorasan province was the most efficient and East Azerbaijan and Tehran ranked second and third. Based on these models, Mazandaran and Qom were the least efficient provinces. Partial frontier models ranked Khorasan and East Azerbaijan as the most efficient cotton producing provinces in the country.

7.2.2 Sugar Beet Production

To model sugar beet technical efficiency, this book used information from 11 sugar beet producing provinces during the period 2000–2012 (Markazi, West Azerbaijan, Kermanshah, Fars, Kerman, Khorasan, Isfahan, Chaharmahal and Bakhtiari, Lorestan, Semnan and Qazvin). Khorasan and West Azerbaijan provinces are the most important provinces for sugar beet. This sample of provinces is different from those used in the study of cotton crop production.

In the case of sugar beet production, the flexible translog functional form was also employed for 12 stochastic frontier models. Pesticides, chemical fertilizers and labor inputs had a positive and statistically significant effect on sugar beet production. Like the results for the cotton crop, chemical fertilizers had the largest partial elasticity among the inputs and it can be concluded that chemical fertilizers played an important role in the production of industrial crops in Iran. Use of more chemical fertilizers as compared to animal fertilizers led to a significant decrease in the technical inefficiency of sugar beet production. The results also showed that an increased use of machinery led to a significant increase in the technical efficiency of sugar beet production in Iran. Highest and lowest average technical efficiency was measured in groups 3 and 2 of the parametric models respectively. According to most of the parametric models, mean technical efficiency of sugar beet producing provinces in Iran was higher than 54%. Based on these models, West Azerbaijan and Kerman were the most and least technically efficient provinces in sugar beet production, respectively. The results of mean technical efficiency during the study period showed that according to most of the time-variant efficiency models there was a decreasing trend in technical efficiency. Like the results for cotton crop, rank order correlation between Models 1, 2, 3 and 11 was high.

Mean technical efficiency measurement based on Window-DEA, Window-FDH, Window Order-m and Window Order-α was 0.67, 0.94, 0.80 and 0.77 respectively. Full frontier models ranked the sugar beet producing provinces in the same order. Based on these models West Azerbaijan, Khorasan and Lorestan were the most efficient sugar beet producing provinces in Iran while Chaharmahal and Bakhtiari and Fars were the least efficient. According to partial frontier models, Kerman and West Azerbaijan provinces were the most efficient and Isfahan was the least efficient sugar beet producing province.

An estimation of technical efficiency in production of cotton and sugar beet crops provided evidence of heterogeneity in production efficiency and the existence of a sufficiently large space for improvements in performance. This information can be used in formulating policies to enhance efficiency through better allocation of public extension service resources and use of inputs at the producer level.

References

Abrishami H., and L. Niakan (2010). "Measuring technical efficiency of Iranian power plants using stochastic frontier analysis (SFA) and comparison with selected developing countries", *Quarterly Energy Economics Review*, 26, 153–175. (In Persian)

Abtahi, Y. and M.R. Islami (2010). "Comparison of provincial productivity efficiency of rain-fed wheat production in Iran", *Journal of Agricultural Extension and Education Researches*, 3(2), 36–25. (In Persian)

Afonso, A. and M. Aubyn (2005). "Non parametric approaches to education and health efficiency in OECD countries", *Journal of Economics*, 3, 227–246.

Afonso, A., L. Schuknecht and V. Tanzi (2003). "Public sector efficiency: an international comparison", Working Paper Series from European Central Bank, No 242.

Afriat, S.N. (1972). "Efficiency estimation of production functions", *International Economics Review*, 13, 568–598.

Ahn, S.C., A.E. Young, Y.H. Lee and P. Schmidt (2007). "Stochastic frontier models with multiple time-varying individual effects", Journal of Productivity Analysis, 27(2), 1–12.

Ahn, S.C., Y.H. Lee and P. Schmidt (2001). "GMM estimation of linear panel data models with time-varying individual effects", *Journal of Econometrics*, 101, 219–255.

Aigner, D.J., C.A.K. Love and P. Schmidt (1977). "Formulation and estimation of stochastic production function models", *Journal of Econometrics*, 6, 21–37.

Ajibefun, I. (2008). "An evaluation of parametric and non-parametric methods of technical efficiency measurement: Application to small scale food crop production in Nigeria", *Journal of Agricultural and Social Sciences*, 4, 95–100.

Alene, A., V. Manyong and J. Gockowski (2006). "The production efficiency of intercropping annual and perennial crops in southern Ethiopia: A comparison of distance functions and production frontiers", *Agricultural Systems*, 91, 51–70.

Amadeh H., A. Emami Meibodi and A. Azadinejad (2009). "Ranking the Iranian provinces by technical efficiency of industrial sector applying DEA method", *Science and Development*, 29, 162–180. (In Persian)

Aragon, Y., A. Daouia and C. Thomas-Agnan (2005). "Nonparametric frontier estimation: a conditional quantile based approach", *Econometric Theory*, 21, 358–389.

Ardabili Mianji, P. and V. Barimnezhad (2016). "Efficiency analysis of agricultural bank using Data Envelopment Analysis (case study of branches in Alborz province)", *Agricultural Economics Reserch*, 8(4), 19–37. (In Persian)

Asmild, M., J.C. Paradi, V. Aggarwall and C. Schaffnit (2004). "Combining DEA window analysis with the Malmquist index approach in a study of the Canadian banking industry", *Journal of Productivity Analysis*, 21, 67–89.

Azar, A. and D. Gholamrezai (2006). "Ranking the provinces of the country using Data Envelopment Analysis approach (Application of Human Development Indicators)", *Iranian Economic Research Journal*, 8(27), 153–173. (In Persian)

© Springer Nature Singapore Pte Ltd. 2018

M. Rashidghalam, *Measurement and Analysis of Performance of Industrial Crop Production: The Case of Iran's Cotton and Sugar Beet Production*, Perspectives on Development in the Middle East and North Africa (MENA) Region, https://doi.org/10.1007/978-981-13-0092-9

Babai M., F. Rastegri and M. Sabuhi Sabuni (2012). "Efficiency evaluation of cucumber greenhouses using Distance Envelopment Analysis method", *Journal of Agricultural Economics and Development*, 26(2), 117–125. (In Persian)

Banker, R.D., A. Charnes and W.W. Cooper (1984). "Some models for estimating technical and scale inefficiency in Data Envelopment Analysis", *Management Science*, 30, 1078–92.

Battese G.E. and S.S. Broca (1997). "Functional forms of stochastic frontier production functions and models for technical inefficiency effects: a comparative study for wheat farmers in Pakistan". *Journal of Productivity Analysis*, 8, 395–414.

Battese, G.E. and T.J. Coelli (1988). "Prediction of firm level technical efficiencies with a generalized frontier production function and panel data", *Journal of Econometrics*, 38(3), 387–399.

Battese, G.E. and T.J. Coelli (1992). "Frontier production functions, technical efficiency and panel data: with application to paddy farmers in India", *Journal of Productivity Analysis*, 3, 153–169.

Battese, G.E. and T.J. Coelli (1995). "A model for technical inefficiency effects in a stochastic frontier production function for panel data", *Empirical Economics*, 20(2), 325–332.

Behrouz, A. and A. Emami Meybodi (2014). "Measurement of technical, allocative, economic efficiencies and productivity of Iran's Agriculture sector using Non-Parametric Method (with emphasis on watermelon production)", *Journal of Agricultural Economics Research*, 6(3), 43–66. (In Persian)

Boles, J.N. (1996). "Efficiency squared- efficient computation of efficiency indexes", Proceedings of the 39th Annual Meeting of the Western Farm Economic Association, 137–142.

Carbone, T.A. (2000). "Measuring efficiency of semiconductor manufacturing operations using data envelopment analysis (DEA)". In IEEE/SEMI advanced semiconductor manufacturing conference, 56–62.

Carvalho, P. and R.C. Marques (2014). "Computing economies of vertical integration, economies of scope and economies of scale using partial frontier nonparametric methods", *European Journal of Operational Research*, 234, 292–307.

Cazals, C., J.P. Florens and L. Simar (2002). "Nonparametric frontier estimation: a robust approach", *Journal of Econometrics*, 106, 1–25.

Chakraborty, K., S. Mirsa and P. Johnson (2002). "Cotton farmers' technical efficiency: stochastic and nonstochastic production function approaches", *Agricultural and Resource Economics Review*, 31(2), 211–220.

Charnes, A., W.W. Cooper and E. Rhodes (1978). "Measuring the efficiency of decision making units", *European Journal of Operations Research*, 2, 429–444.

Charnes, A., W.W. Cooper, A.Y. Lewin and L.M. Seiford (1994). Data Envelopment Analysis: Theory, Methodology, and Application, Kluwer Academic Publishers, Norwell.

Chiami, B.C. (2011). Determinants of technical efficiency in smallholder sorghum farming in Zambia. MSc Thesis, Graduate School of the Ohio State University.

Coelli, T.J. (2008). A guide to DEAP version 2.1: A data envelopment analysis (computer) program, CEPA Working Paper 96/08.

Collier, T., A. Johnson and J. Ruggiero (2011). "Technical efficiency estimation with multiple inputs and multiple outputs using regression analysis", *European Journal of Operations Research*, 208, 153–160.

Colombi, R., G. Martini and G. Vittadini (2011). A stochastic Frontier Model with Short-run and Long-run Inefficiency Random Effects. Department of Economics and Technology Management, Universita di Bergamo, Italy.

Colombi, R., S.C. Khumbakhar, G. Martini and G. Vittadini (2014). "Closed-Skew Normality in stochastic frontiers with individual effects and long/short-run efficiency". *Journal of Productivity Analysis*, 42(2), 123–136.

Danijela, P., A. Radonjic, Z. Hrle and V. Čolic (2012). "DEA window analysis for measuring port efficiencies in Serbia", *Traffic and Transportation*, 24(1), 63–72.

Daouia, A. and L. Simar (2007). "Nonparametric efficiency analysis: a multivariate conditional quantile approach", *Journal of Econometric*, 140(2), 375–400.

Daouia, A. and L. Simar (2005). "Robust nonparametric estimators of monotone boundaries", *Journal of Multivariate Analysis*, 96, 311–331.

Daraio, C., and L. Simar (2007). Advanced Robust and Nonparametric Methods in Efficiency Analysis: Methodology and Applications, New York: Springer.

Dashti, G., S. Yavari, A. Pishbahar and B. Hayati (2012). "Effective factors on technical efficiency of broiler breeding units in Sonqor and Kalayi county", *Journal of Animal Science Research*, 21(3), 83–95. (In Persian)

De witte, K. and R.C. Marques (2010). "Influential observations in frontier models, a robust non-oriented approach to the water sector", *Annals of Operational Research*, 181(1), 377–392.

DeBorger, B., K. Kerstens, W. Moesen and J. Vanneste (1994). "A non-parametric free disposal hull (FDH) approach to technical efficiency: An illustration of radial and graph efficiency measures and some sensitivity results", *Swiss Journal of Economics and Statistics*, 130(4), 647–667.

Deprins, D. and H. Tulkens (1984). "Measuring labour efficiency in post offices". In Marchand, M. and Tulkens, H. (eds.) the Performance of Public Enterprises: Concepts and Measurement North-Holland, 243–267.

Desli, E., S.C. Ray and S.C. Kumbhakar (2003). "A dynamic stochastic frontier production model with time-varying efficiency", *Applied Economics Letters*, 10, 623–626.

Durandish A., M. Kohansal, N. Shahnushi Furushani, M. Hosseinzadeh (2012). "Evaluation of technical efficiency of barberry producers in Southern Khorasan province", *Journal of Agricultural Economics*, 6(2), 101–120. (In Persian)

Emokaro, C. and P. Ekunwe (2009). "Efficiency of resource-use and elasticity of production among catfish farmers in Kaduna, Nigeria", *Journal of Applied Sciences Research*, 5(7), 776–779.

Emvalomatis, G. (2009). Parametric Models for Dynamic Efficiency Measurement, Doctoral dissertation, Department of Agricultural Economics and Rural Sociology, The Pennsylvania State University.

Fare, R. and C.A.K. Lovell (1978). "Measuring the technical efficiency of production", *Journal of Economic Theory*, 19, 150–162.

Farrell, M.J. (1957). "The Measurement of productive efficiency", Journal of the Royal Statistical Society. Series A (General), 120(3), 253–290.

Fried, H.O., C.A.K. Lovell, S.S. Schmidt (2008). The Measurement of Productive Efficiency and Productivity Growth. Oxford University Press, New York.

Gabdo, B.H., I.B. Abdlatif, Z.A. Mohammed and M.N. Shamsuddin (2014). "Comparative estimation of technical efficiency in livestock-oil palm integration in Johor, Malaysia: evidence from full and partial frontier estimator", *Journal of Agricultural Science*, 6(3), 140–150.

Goyal, K. and S. Suhag (2003). Estimation of technical efficiency on wheat farms in northern India- a panel data analysis. International farm management congress.

Greene, W.H. (1980). "On the estimation of a flexible frontier production model", *Journal of Econometrics*, 13(1), 101–115.

Greene, W.H. (1990). "A gamma distributed stochastic frontier model", *Journal of Econometrics*, 46(1), 141–164.

Greene, W.H. (2005a). "Fixed and random effects in stochastic frontier models". *Journal of Productivity Analysis*, 23, 7–32.

Greene, W.H. (2005b). "Reconsidering heterogeneity in panel data estimators of the stochastic frontier model", *Journal of Econometrics*, 126, 269–303.

Gul, M., B. Koc, E. Dagistan, M.G. Akpinar and O. Parlakay (2009). "Determination of technical efficiency in cotton growing farms in Turkey: A case study of Cukurova region", *African Journal of Agricultural Research*, 4(10), 944–949.

Haeri, A and A. Asayesh (2009). "Analysis the status of cotton in Iran and the world, Bureau of Statistical and Strategic Studies of the textile industry of Iran". (In Persian)

Hakimipour, N. and K. Hojabrkiani (2008). "Comparative analysis of the efficiency of large industries in Iranian provinces: using Stochastic Frontier Function", *Journal of Science and Research (Scientific-Research)*, 15(24), 138–167. (In Persian)

Hallam, D. and F. Machado (1995). "Efficiency analysis with panel data: A study of Portuguese dairy farms", *European Review of Agricultural Economics*, 23, 79–93.

Hardeman, S. and V.V. Roy (2013). "An analysis of national research systems (II): Efficiency in the production of research excellence". Joint Research Centre - European Commission; Econometrics and Applied Statistics Unit - Institute for the Protection and Security of the Citizen. Protection and Security of the Citizen - Via E. Fermi 2749, TP 361 - Ispra (VA), I-21027, Italy.

Henderson, D.J. (2003). The Measurement of Technical Efficiency Using Panel Data. Department of Economics, State University of New York at Binghamton, May.

Hoseinpour, A., R. Moghadasi and S. Yazdani (2014). "Study of technical efficiency and its determinants on Kashan's Rosewater Industry", *Iranian Journal of Medicinal and Aromatic Plants*, 30(1), 42–56. (In Persian)

Jondrow, J., C.A.K. Lovell, I.S. Materov and P. Schmidt (1982). "On the estimation of technical inefficiency in the stochastic frontier production function model", *Journal of Econometrics*, 19 (2-3), 233–238.

Kamruzzaman, M., and M. Islam (2008). "Technical efficiency of wheat growers in some selected sites of Dinajpur district of Bangladesh", *Bangladesh Journal of Agriculture Research*, 33(3), 363–373.

Karagiannis, G. and V. Tzouvelekas (2009). "Parametric measurement of time-varying technical inefficiency: results from competing models", *Agricultural Economics Review*, 10(1), 50–79.

Karami, A., S.F. Eftekhari and A. Abdshahi (2012). "The efficiency evaluation of Early Burden Companies in Kohghiluyeh and Boyer Ahmad province (milk cow, chicken and fish farming)", *Journal of Agricultural Economics*, 4(3), 59–76. (In Persian)

Khajepour, M. (1373). Production of Industrial Plants, Jihad publishing house of Isfahan University, Isfahan, Iran.

Kumbhakar, S.C. (1987). "Production frontiers and panel data: an application to U.S. class 1 railroads", *Journal of Business and Economics Statistics*, 5, 249–255.

Kumbhakar, S.C. (1990). "Production frontiers, panel data, and time-varying technical inefficiency", *Journal of Econometrics*, 46, 201–212.

Kumbhakar, S.C. (1991). "Estimation of technical inefficiency in panel data models with firm- and time-specific effects", *Economic Letters*, 36, 43–48.

Kumbhakar, S.C. and A. Heshmati (1995). "Efficiency measurement in Swedish dairy farms: an application of rotating panel data, 1976-88", *American Journal of Agricultural economics*, 77, 660–674.

Kumbhakar, S.C. and C.A.K. Lovell (2000). Stochastic Frontier Analysis. Cambridge University Press, Cambridge.

Kumbhakar, S.C. and H.J. Wang (2005). "Estimation of growth convergence using a stochastic production function approach", *Economic Letters*, 88, 300–305.

Kumbhakar, S.C. and L. Hjalmarsson (1993). "Technical efficiency and technical progress in Swedish dairy farm". In: Fried HO, Lovell CAK, Schmidt SS (eds) The measurement of productive efficiency- techniques and applications. Oxford University Press, Oxford, 256–270.

Kumbhakar, S.C. and L. Hjalmarsson (1995). "Labor- use efficiency in Swedish social insurance offices", *Journal of Applied Economics*, 10, 33–47.

Kumbhakar, S.C., G. Lien and J.H. Hardaker (2014). "Technical efficiency in competing panel data models: a study of Norwegian grain farming", *Journal of Productivity Analysis*, 14(2), 321–337.

Kumbhkar, S.C., H.J. Wang and A.P. Horncastle (2015). A Practitioner's Guide to ha Stochastic Frontier Analysis Using Stata. Cambridge University Press, New York.

Lambarraa, F. (2012). "The Spanish Horticulture Sector: A dynamic efficiency analysis of Outdoor and Greenhouse farms", Selected Paper prepared for presentation at the International Association of Agricultural Economists (IAAE) Triennial Conference, Foz do Iguaçu, Brazil, 18–24 August, 2012.

Lee, Y.H. and P. Schmidt (1993). "A production frontier model with flexible temporal variation in technical efficiency". Chapter 8, in the measurement of productive efficiency techniques and applications, eds., Fried, H., C.A.K.

Lee, Y.H. and P. Schmidt (1995). "GMM estimation of a panel data regression model with time-varying individual effects". Unpublished manuscript, Arizona State University.

Meeusen, W. and J. Van Den Broeck (1977). "Efficiency estimation from Cobb-Douglas production functions with composed error", International Economic Review, 18(2), 435–444.

Ministry of Agriculture Jihad (2013). Agricultural statistics, Department of economic and planning. Center for Information and Communication Technology.

Ministry of Agriculture Jihad (2015). Agricultural statistics, Department of economic and planning. Center for Information and Communication Technology.

Ministry of Agriculture Jihad (2017). Agricultural statistics, Department of economic and planning. Center for Information and Communication Technology.

Mohammed, R. and S. Saghaian (2014). "Technical efficiency estimation of rice production in South Korea". Selected paper prepared for presentation at the 2014 Southern Agricultural Economics Association (SAEA) Annual Meetings in Dallas.

Mojarad, A., A.A. Kahkha and M. Sabuhi Saboni (2011). "Introduction of a stochastic nonparametric method in estimating technical efficiency: (case study of poultry farms in Sistan region)", Journal of Agricultural Economics, 3(3), 91–106. (In Persian)

Motafakerazad, M.A., M. Pour Abadollah Kuich, F. Fallahi, R. Rangpour and S. Sojudi (2014). "Calculation of technical efficiency of Iranian thermal power plants and investigating its determinants: application of stochastic nonparametric data envelopment analysis. Journal of Economic Research, 49(1), 93–113. (In Persian)

Mundlak, Y. (1961). "Aggregation over time in distributed lag models", International Economic Review, 2, 154–163.

Pishbahar, A. and A. Nasiri (2012). "Evaluation of technical efficiency of strawberry producers in Sanandaj". Journal of Agricultural Science and Sustainable Production, 22(4), 125–134. (In Persian)

Pitt, M. and L.F. Lee (1981). "The measurement and sources of technical inefficiency in the Indonesian weaving industry", Journal of Development Economics, 9, 43–64.

Pjevcevic, D., A. Radonjic, Z. Hrle and V. Colic (2012). "DEA Window Analysis for measuring port efficiencies in Serbia", Traffic and Transportation, 24(1), 63–72.

Rafati, M., Y. Azarinfar, M. Zad, A. Barabari and M. Kazemnejd (2011). "Study of technical, allocative and economic efficiency of cotton producers in Golestan province using parametric method (case study of Gorgan county)", Journal of Agricultural Research, 3(1), 121–142. (In Persian)

Rashidghalam, M. (2008). Effects of removing agricultural input subsidies on sugar beet production of Iran. Master thesis, Faculty of Agriculture, Tarbiat Modares University. (In Persian)

Rashidghalam, R., A. Heshmati, G. Dashti and E. Pishbahar (2016). "Comparison of Panel Data Models in estimating Technical Efficiency" IZA Discussion Paper 2016:9807, and CESIS Electronic Working Paper 2016:433, 31 pages.

Rasouli, F. (2011). Determining energy indicators in sugar beet farms and assessing their efficiency using Data Envelopment Analysis (case study of Mahabad county), Master Thesis, University of Tabriz. (In Persian)

Řepková, I. (2014). "Efficiency of the Czech banking sector employing the DEA window analysis approach", Procedia Economics and Finance, 12, 587–596.

Richmond, J. (1974). "Estimating the efficiency of production", International Economic Review, 15(2), 515–521.

Ross, A. and C. Droge (2002). "An integrated benchmarking approach to distribution center performance using DEA modeling", Journal of Operations Management, 20, 19–32.

Ruggiero, J. (2007). "A comparison of DEA and the stochastic frontier model using panel data", International Transactions in Operational Research, 14, 259–266.

Sakhanvar, M, H. Sadeghi, A. Assari, K. Yavari and N. Mehregan (2011). "Application of Window Data Envelopment Analysis to study the Performance of Iranian Power Distribution Companies", *Economic Growth and Development Research*, 1(4), 145–182. (In Persian)

Schmidt, P. and C.A.K. Lovell (1979). "Estimating technical and allocative inefficiency relative to stochastic production and cost frontiers", *Journal of Econometrics*, 9, 343–366.

Schmidt, P. and R.C. Sickles (1984). "Production frontiers and panel data", *Journal of Business and Economic Statistics*, 4, 367–374.

Seydan, M. (2004). "Assessment of technical inefficiency determinants of garlic producers: case study of Hamadan province", *Journal of Research and Development in Agriculture and Horticulture*, 64, 74–79. (In Persian)

Seyyed Sharifi R. (2013). Industrial Crops. Amidi, University of Mohaghegh Ardebili, Ardebil, Iran. (In Persian)

Shafiee, L. (2007). "Determination of technical, allocative and economic efficiency of sugar beet producers in Bardsir county", *Sugar beet journal*, 22(2), 109–121. (In Persian)

Shepherd, R.W. (1970). Theory of Cost and Production Functions, Princeton, Princeton University Press.

Silva, T.D., C. Martins-filho and E. Ribeiro (2016). "A comparison of nonparametric efficiency estimators: DEA, FDH, DEAC, FDHC, order-m and quantile", *Economics Bulletin*, 36 (1), 118–131.

Simar, L. and P.W. Wilson (2000). "Statistical inference in nonparametric frontier models: recent developments and perspectives", *Journal of Productivity Analysis*, 13, 49–78.

Sokhanvar, M. (2011). "Window data envelopment analysis for evaluating technical efficiency trend of Iranian Power Distribution Companies", *Quarterly Journal of Economic Growth and Development*, 1(4), 146–182. (In Persian)

Solís, D., B. Bravo-Ureta and R. Quiroga (2009). "Technical efficiency among peasant farmers participating in natural resource management programmes in Central America", *Journal of Agricultural Economics*, 60, 202–219.

Song, W., Z. Hen and X. Deng (2016). "Changes in productivity, efficiency and technology of China's crop production under rural restructuring", *Journal of Rural studies*, 47(B), 563–576.

Stevenson, R.E. (1980). "Likelihood functions for generalized stochastic frontier estimation", *Journal of Econometrics*, 13(1), 57–66.

Subhash, C.R. (2004). Data Envelopment Analysis: Theory and Techniques for Economics and Operations Research. Cambridge University Press, UK, New York

Sueyoshi, T and S. Aoki (2001). "A use of a nonparametric statistic for DEA frontier shift: The Kruskal and Wallis rank test", *OMEGA*, 29, 1–18.

Tauchmann, H. (2011). "Partial frontier efficiency analysis for Stata", Discussion paper, SF 823.

Wang, H.J. and C.W. Ho (2010). "Estimating fixed-effect panel data stochastic frontier models by model transformation", *Journal of Econometrics*, 157(2), 286–296.

Wang, K., Y. Shiwei and W. Zhang (2013). "China's regional energy and environmental efficiency: A DEA window analysis based dynamic evaluation", *Mathematical and Computer Modelling*, 58, 1117–1127.

Wang, Y. and Q. Liu (2006). "Comparison of Akaike information criterion (AIC) and Bayesian information criterion (BIC) in selection of stock–recruitment relationships", *Fisheries Research*, 77, 220–225.

Wilson, P.W. (1993). "Detecting outliers in deterministic nonparametric frontier models with multiple outputs", *Journal of Business and Economic Statistics*, 11, 319–323.

Yadollahi, H. (2003). Evaluation and estimation of efficiency and productivity in Iranian Industries, Master Thesis, Allameh Tabatabaei University. (In Persian)

Yaghubi M., J. Shahraki and A. Karbasi (2010). "Efficiency survey of cooperative and non-cooperative shrimp razing units in Chaharmahal and Bakhtiari county using Data Envelopment Analysis (application of CCR and FDH models)", *Cooperation Journal*, 21(4), 71–95. (In Persian)

Yang, H.H. and C.Y. Chang (2009). "Using DEA window analysis to measure efficiencies of Taiwan's integrated telecommunication firms", *Telecommunications Policy*, 33, 98–108.

Yazdani, S. and R. Rahimi (2012). "Evaluation of the efficiency of sugar beet production in Qazvin plain", *Journal of Sugar Beet*, 28(2), 113–118. (In Persian)

Zeranejad, M., and R. Yousefi (2009). "Evaluation of technical efficiency of wheat production in Iran using parametric and nonparametric approaches", *Economic Research*, 9(2), 172–145. (In Persian)

Zeranejad, M., F. Khodadad Kashi and R. Yousefi Haji Abad (2012). "Evaluation of technical efficiency of Iranian Industries", *Quarterly Journal of Quantitative Economics*, 9(2), 31–48. (In Persian)

Zhang, X.P., X.M. Cheng, J.H. Yuan and X.J. Gao (2011). "Total-factor energy efficiency in developing countries", *Energy Policy*, 39, 644–650.